1986

BY THE SAME AUTHOR

*Gifts of Unknown Things*
*Supernature*
*The Romeo Error*

# LYALL WATSON

SIMON AND SCHUSTER     NEW YORK

# LIFETIDE

The Biology of the Unconscious

Published by Simon & Schuster
A Division of Gulf & Western Corporation
Simon & Schuster Building
Rockefeller Center
1230 Avenue of the Americas
New York, New York 10020
Designed by Edith Fowler
Manufactured in the United States of America

1 2 3 4 5 6 7 8 9 10


Library of Congress Cataloging in Publication Data

Watson, Lyall.
  Lifetide: the biology of the unconscious.

  Bibliography: p.
  1. Consciousness—Physiological aspects. 2. Biology—Philosophy. 3. Biology—Miscellanea. I. Title.
BF311.W29    001.9    78-24305
ISBN    0-671-24689-5

# Contents

The collective unconscious is common to all; it is the foundation of what the ancients called the "sympathy of all things."

—CARL GUSTAV JUNG,
*Memories, Dreams, Reflections*

# LIFETIDE

# THE TIDEWAY: A Preview

This is not a time for certainty. We seem in recent years to have grown through the confident adolescence of science into a philosophical maturity, prepared not only to admit our ignorance, but to come to terms with the fact that there are some things we can never know. And that it doesn't matter. Armed only with the Principle of Uncertainty and a host of hidden variables, we seem to be better equipped than ever before to break through some of the misty fringes on the edge of the unknown. Not in search of knowledge, for we can now see that was a kind of conceit, but in the humble hope of more clearly defining our area of understanding.

It seems appropriate too that in these new endeavors, it is the most heavily funded programs, those sponsored by purely practical and military concerns and involving the most complex machinery in elaborate experiments of great scientific precision, that are turning out the most simple poetic truths. They end, in the words of one physicist, in the realization that "we can never speak about nature without, at the same time, speaking about ourselves."[99]*

We are beginning to feel involved. To experience the universe and our participation in it as one dynamic, inseparable whole; an organism, ever-moving and alive; spiritual and material all at the same time.

I glory in this, and regret only that word of this new awareness is not filtering down to all levels as fast as it deserves. With a few joyous exceptions like Fred Hoyle and Carl Sagan, individual scientists are a sober, industrious lot, going on with their research in a careful, methodical manner and giving way only occasionally, usually in bouts of officially sanctioned levity at end-of-conference parties, to a certain amount of restrained speculation; when what some of the wonders revealed to them merit is that they should come rushing out into the world hallooing and giving out the news. Eventually it may all become too much for them and they will give way in

*Superior numbers throughout refer to numbered items in the bibliography.*

awe and admiration; but until such time as scientists take again to running through the streets of Syracuse with glad cries, the responsibility rests on intermediaries like myself, who are privy to some of the secrets and able to read between the carefully coded lines.

This is a heavy responsibility, a trust not to be taken lightly, or abused by selecting only those bits of news which best support some political conviction or private persuasion. But, at the same time, I cannot view the facts dispassionately, for I too am intimately involved, and can only hope that my interpretation of them does justice both to the seekers and to the search.

So this is a hopelessly ambitious book about the universe, and about us. It represents a synthesis of some recent revelations and a statement of what I have come to believe as a result of a long period of exposure both to nature and to naturalists. It deals with general principles, not with "laws of nature," because there is reason to doubt that any such invariables exist. And it is concerned largely with forces that operate through, and may have been responsible for, our current consciousness. Much of the subject matter is what is usually known as the supernatural—because I hold with Carl Gustav Jung that the only part of us in close touch with our roots is the unconscious, and that in normal people this manifests itself most clearly in so-called occult phenomena. There is much to be learned about nature through techniques evolved in all the life sciences, but the real insights seem to me to lie in anomalous experience, in loose ends on the fringes of understanding. The truth, or our best approximation of it, is carried somewhere in the undercurrents of life—and the tide is coming in.

Sigmund Freud, in a moment of weakness when he and Jung were on the verge of separation, portrayed the occult as an invidious "black tide."[301] It disturbed him, as it still does many people today; but I feel that the ominous aspects of the supernatural are due largely to misunderstandings and that the label "black" is misleading and unnecessary. I do, however, like the tidal metaphor and intend to apply it to the whole panoply of hidden forces that shape life in all its miraculous guises. These are the eddies and vortices of nature that flow together to form the living stream. They are the substance of the Lifetide.

The classical psychoanalytic interpretation of the image of a rising tide in a dream would be that it indicates an emotion on the increase. If we enlarge the tidal metaphor and see the mind as an ocean lapping at a physical shore, then the periodic rise and fall of interest in the occult can probably be said to be now near high water, perhaps even at spring-tide level. Jung saw it in astrological terms as a manifestation of "psychic changes which always appear at the end of one Platonic month and at the beginning of another."[300] A sort of changing of the guards by the gods, which takes place wherever there is a long-lasting transformation of the collective psyche.

I believe we are now at such a point in time—not because of any mystical signs or portents, but because we have reached a critical stage in our understanding of reality and of ourselves, a sort of evolutionary embankment which seems only just strong enough to hold the Lifetide at bay. Either the waters recede, probably to return again in the next cycle, or something stirs them up at the vital moment, sending waves crashing over the retaining walls in a flood which could change the face of the land forever.

Charles Darwin gave rise to such a wave; Freud produced another; and the effects of both are still being felt. But I suspect that what might actually carry the waters over the top this time is more of a ground swell, a general shift in awareness produced by deep undulations in the human spirit. And perhaps because I happen to be a biologist, and therefore particularly sensitive to the ripples in my own part of the pool, I see particular significances in those new insights which help to put the problem into a more meaningful evolutionary perspective.

When examined closely, these trends combine to produce a picture of ourselves and our origins quite different from the one we currently hold to be true. This new view may be as distorted as the old one, and due in its turn to be replaced in the same way, but while it lasts it has the salutary effect of forcing us to think again—to reappraise some of our most fundamental assumptions about life, and to realize that each time we adopt a confident scientific stance, it is one with only temporary validity, based on several questionable articles of faith. We are, even in our most materialistic moments, strangely religious animals.

We take an incredible amount on trust. In an increasingly complex world, we have to. It is difficult even for an expert to keep up with the literature in his or her own field, let alone keep a watchful eye on others. But it would be as well if we appreciated this fact; if we admitted to ourselves that we structure our environment for convenience, and that our current view of reality is little more than a working hypothesis, accepted for the moment by consensus—and susceptible to rejection in precisely the same way. We still have that power, and little by little, often unwittingly, we continue to exercise it.

People keep stumbling over things, finding answers to questions that hadn't even consciously been asked. Recent discoveries in cell biology and genetics, for instance, reveal an astonishing autonomy at that level, a strange molecular freedom of action which makes a mess of all mechanistic theory. It is a wonderful mess because it demonstrates, just when we needed reminding, that we and our life system are very much more extraordinary than many will allow. The everyday commonplace things that we all take for granted, and think we understand, turn out to be elusive and mysterious. It may come as a surprise to learn that, despite years of intensive and expensive research, nobody yet knows the answer to even the most basic questions, such as what makes us sleep, or why we tend to forget our dreams, or how we ever remember anything at all. Each time a line of research seems to be on the brink of a definitive discovery, on the point of putting an end to centuries of speculation, the last in an ever-diminishing nest of puzzle boxes opens unexpectedly to reveal yet another whole universe beyond. Everything around us is filled with mystery and magic.

I find in this no cause for despair, no reason to turn for solace to esoteric formulae or chariots of gods. On the contrary, our inability to find easy answers fills me with a fierce pride in our ambivalent biology, with a determination to try to rephrase the old questions in a new and more meaningful way, and with a constant sense of wonder and delight that we should be part of anything so profound.

Lyall Watson
Oxford, England, 1978

# PART ONE

# The Brink of the World

I had the feeling that I had pushed to the brink of
the world; what was of burning interest to me was
null and void for others, and even a cause for
dread.
   Dread of what? I could find no explanation for
this. After all, there was nothing preposterous or
world-shaking in the idea that there might be
events which overstepped the limited categories of
space, time, and causality.

—CARL GUSTAV JUNG,
*Memories, Dreams, Reflections*

I have always felt that Venice was unreal.

There is something inconsistent, a transiency like that of an image in a dream, about the combination of rich brocade and crumbling facade, soft light and dark water, preposterous eighteenth-century furniture and ultramodern glass. But the features which disturb me most are those on the carved heads which glare down from every bridge and building. Wall-eyed gorgons and giants with tusks and pendulous tongues. Grotesque, hairless heads wrinkled in enjoyment of outdated jokes. Death's heads, carnival masks. And, on the wall above a first-story window not far from Santa Maria dei Miracoli, the tearful face of a child.

Across the *calle*, in direct sight of that head, live an unusual family: unusual for Italians in that they have only one child; and unusual by any standards in that this little girl, at the age of five, began to contradict everything we know about space, time, and causality.

I may have been partly to blame. The Italian translation of my first book about the occult was published in May of 1974. Late that autumn the girl's father, who works in one of Venice's luxury hotels, bought a copy of *Supernatura* to read on the *vaporetto* as he traveled across to the Lido early each morning to play tennis there with other off-duty friends. The season and the tennis ended, but the game had just begun. And when it did, when five-year-old Claudia invented something absolutely unique, rather than take the matter to his priest, her father wrote directly to me in a mixture of pride and panic I found impossible to ignore.

The family lives in a three-room, high-ceilinged apartment, sparsely furnished but simmering with the smells of good cooking. Dallying over a dish of fish and *polenta*, I had the chance to watch Claudia, and she the time to get used to me. She was small even for her age, with wispy hair, tiny and totally self-controlled, preternaturally still for a child. When her hands were not actually doing some necessary thing, they lay in her lap or on the table, at rest. She seemed to live through her eyes, which were enormous, black and disconcertingly knowing. They probed me, dissected me, sifted through the components, filed the useful bits away for future reference, and then turned inward again to the things that really mattered.

After dinner her father and I sat and talked while Claudia paged through a magazine. Then, very casually, he opened a tube of tennis balls that stood on a corner table and rolled one across the carpet so that it came to rest right on the picture she was examining. She favored him with one of her discerning looks and, almost in resignation, set the *rivista* aside and turned her attention to the ball. She held it to her cheek, affectionately, and then balanced the ball on her left hand while she stroked it gently with her right as though it were a small furry animal, a dormouse to be roused from untimely hibernation. It was a pretty scene, an arresting portrait of the hopes and fears of youth just as Lorenzo Lotto captured them four hundred years before in that very neighborhood. But my appreciation was cut short, and I hurtled back to the present in total terrified incomprehension, when the dormouse broke all the rules and responded.

One moment there was a tennis ball—the familiar off-white carpeted sphere marked only by its usual meandering seam. Then it was no longer so. There was a short implosive sound, very soft, like a cork being drawn in the dark, and Claudia held in her hand something completely different: a smooth, dark, rubbery globe with only a suggestion of the old pattern on its surface—a sort of negative, through-the-looking-glass impression of a tennis ball.

Claudia seemed not to be surprised, perhaps a bit pleased, as she handed the transformed ball back to her father, who passed it on to me. I wasn't at all sure I wanted anything to do with it, until I realized what it was. It was something I had never seen before, but recognized instantly despite the unfamiliar point of view. It wasn't a bald tennis ball, deprived somehow of its hair, but an everted tennis ball, one turned inside out yet still containing a volume of air under pressure. I squeezed it and it held. I dropped it and it bounced. I picked up a knife from the dinner table and, with some difficulty, pierced the rubber and let the air hiss out. Then I cut right around the circumference and there it was, lining the interior where it had no business being, the usual furry pile apparently none the worse for wear.

Later that evening, with some reluctance, Claudia did it again and I carried the talisman of an intact everted tennis ball off to my hotel. For two days it sat there on my baroque mantelpiece like a mandala, unmoving but nevertheless mocking me. A sphere, the classic symbol of totality and order, the very shape of the soul, but this one transformed by the child and transfigured by the knowledge that order had been interfered with, that nothing was quite what it seemed.

It still disturbs me. I know enough of physics to appreciate that you cannot turn an unbroken sphere inside out like a glove. Not in this reality.

There is a branch of geometry called topology, which deals with the way in which figures are connected rather than with their shape and size. It is concerned with those factors which remain unchanged when an object undergoes a continuous deformation by bending, stretching, or twisting. And in topological analysis it is possible to imagine a mathematical sphere, made up of collections of points, that can be everted by pushing one side of the ball through the other. But even in such a theoretical exercise, if you obey the basic rule that the deformation must be produced without breaking or tearing the fabric, you are still left with a looped ridge along one surface of the ball.

Only in the last few years has a French mathematician succeeded in proving that there is one indirect way in which a hypothetical, infinitely elastic, hollow sphere can be everted. His procedure involves torturing the ball through a dozen weird, saddlelike shapes whose sides pass through each other several times. With close attention and a vivid imagination you can just, and only just, envisage the steps in the process; but it is obvious that it has no parallel in what we regard as objective reality, and could never have been invented by anyone thinking in pictures rather than mathematical symbols. And there is an intrinsic rightness, a sort of natural poetic justice in the fact that the man who solved the problem of everting a sphere in this way is blind.[384]

And so my un-tennis ball has become for me a sort of symbol, the manifestation of a new, an alternative approach to life—another way of looking at things.

I have given up wondering about the mechanism and no longer itch to get Claudia attached to an instrument for measuring microwave emission or the electrical resistance of the skin. I have learned anyway, from research on the growing band who bend metal and move pendulums in apparently paranormal ways, that it is seldom possible to establish very much more than that these things do happen, sometimes, usually when you least expect it, but seldom in a way that will carry sufficient authority to convince anyone who requires very rigid mechanical demonstrations of reality.

It is impossible to prove, in the normal scientific way, that such things do or don't happen. One is forced to take uncomfortable refuge in the notion that there are other realities, some of them far too delicate and mysterious for totally objective common sense. These systems have a way of transcending ordinary logic and language, which never seem to go quite far enough. Werner Heisenberg, who formulated the Uncertainty Principle, said that "every word or concept, clear as it may seem to be, has only a limited range of applicability."[246] We find that hard to accept and, because our well-worn representation of reality, the one we happen to be wearing with greatest comfort at the moment, is so much easier to deal with than reality itself, we tend to confuse the two. We take our myths and symbols for the real thing.

The greatest difficulty in which the scientific method has landed us is its implicit assumption that observers and experimenters are external to and independent of the objects of their attention. There is good reason to doubt that this is, or ever was, true. Quantum physics is quite clear on the matter. Wanting something, it says, inevitably changes the thing you want. To prove that the tennis ball "really" has been everted, you must cut it open and destroy its very nature. Then all you are left with are shells that quickly revert to the conformation expected of them; to that description of reality which most of us have learned to make in common and to accept as exclusive fact.

But for anyone touched by magic, as I was in Venice, things can never be quite the same again.

I treasure the transformation and I try to find ways of making it work, of fitting it into an evolutionary view in which concepts of causation and purpose are not totally irrelevant. But it isn't easy. The search for validity through proof is fundamentally foreign to magic. I find myself agreeing with the Trobriand Islanders who assume that the importance of a spell lies not in its results, not in proof, but in its very being. "In the appropriateness of its inheritance, in its place within the patterned activity, in its being performed by the appropriate person, in its realization of its mythical basis."[340]

I find it helps me to lose some of my illusory certainty if I close my eyes a little. It was possible for Newton to be confident that "facts" had a stable eternity outside the contaminating range of the human mind, but we can see further now and can't afford to be that dogmatic. It is becoming clear that to observe things is to alter them, and to define and understand anything is tantamount to changing it beyond all recognition.

So the first step in a new approach has to be a different, less obtrusive method of observation.

"If you paint," said Picasso, "close your eyes and sing. Painters should have their eyes put out like canaries, so they'll sing better."[333] Hans Arp wrote: "Under lowered lids, the inner movement streams untainted to the hand. In a darkened room it is easier to follow guidance than in the open air. A conductor of inner music, the great designer of prehistoric images, worked with eyes turned inwards. So his drawings gain in transparency; are open to penetration, to sudden inspiration, to recovery of the inner melody, to the circling approach; and the whole is transmuted into one great exaltation."[101]

I see the sense in the circling approach, the sideways glance, and feel certain that if we are to break the impasse that exists between old science and new needs, we will have to look at things in a different way, as blind men and artists always have.

In attempting to portray a whole subject on a flat surface, most Western artists in literate societies—where everyone has a basic, left-to-right, top-to-bottom, linear visual bias—employ a three-dimensional perspective. They show only what it looks like from a single point in space at a single moment in time. In short, they fail. But preliterate artists, untied to any visual convention, show the subject in full face and profile, from above and below, back and front, inside and out, all simultaneously. A bear in Indian British Columbia is represented complete with skin, bones, entrails, and environment all taken apart and reconstructed on a flat surface to retain every significant element of the whole animal. A fish in aboriginal Australia is seen with X-ray eyes that lay bare its entire being.

New waves in Western art have learned these valuable lessons. Now it is time for Western science to rid itself of similar restrictive and outdated conventions; time we began to build a tissue bank of subjective experience to set alongside our wealth of objective experiment; time we imbued the body of science with real life by giving it a mind.

In this first section, I intend to concentrate more on the pattern than the parts. I want to try to look right through the water and the waves in an attempt to see the tide itself, or at least the mechanism which drives it.

# CHAPTER ONE

# THE SEED: Origins in Space

The sun rises. In that short phrase, in a single fact, is enough information to keep biology, physics, and philosophy busy for all the rest of time.

It means, just for a start, that the sun shines. That four million tons of matter are destroyed at its surface every second in a raging nuclear storm which buffets us here on the fringes of the solar atmosphere. And all the time, the mass of our star is being reduced and its volume and density are changing. It and all other stars are evolving, because evolution is a cosmic operation.

Sunrise also means that earth moves. In addition to the planet's daily rotation and its yearly revolution about our local star, there is now evidence for a much grander motion. It seems that the earth moves with respect to the universe as a whole, and that even this is by no means static.[19]

In 1845 a Dutch meteorologist conducted a splendid experiment. He hired an orchestra of trumpeters and persuaded them to play a fanfare while standing on the flatbed of an open coach in a railroad train hurtling through the countryside outside Utrecht. Buys Ballot and his friends enjoyed the performance from a platform beside the tracks and found, as he predicted, that the sounding brass was higher-pitched when the train was coming toward them than it was when the musicians disappeared in the opposite direction.[575]

You can test the effect yourself, at considerably less expense, by listening to the engine sound of a truck on any modern motorway. The change in frequency as it passes you is due to the Doppler Effect, and this applies to light as well as sound.

If a bright white object is traveling toward you at sufficient speed, its light will be distorted to waves of shorter length, which will give it a more or less violet appearance. As it travels away from you, the wavelengths will increase and it will seem to be more red in color. The extent of this illusion, now known as the "red shift," can be used to calculate how fast the object is moving with respect to your position. And using it, astronomers have been able to show that aside from a few close neighbors, all the other galaxies in the universe are rushing away from us at speeds of many millions of miles an hour.

This does not mean that there is anything special or repulsive about our piece of space. The same thing appears to be happening everywhere. As near as we can tell, every galaxy is rushing away from every other galaxy as though the universe were undergoing some sort of gigantic explosion. This interpretation first appeared soon after 1929 when the American astronomer Edwin Hubble showed that the speed with which other galaxies receded was directly proportional to their distance from us. This means that at some vast distance, remote galaxies must reach the speed of light and vanish from our sight. This distance, the Hubble Radius, effectively limits the size of that portion of the universe which we can ever come to know. It is a sphere with a radius of about thirteen billion light-years. For the sake of some sort of clarity, let's call this area the cosmos.

If the contents of the cosmos are indeed rushing apart, then they must once have been closer together. And, knowing their relative speeds and distances, it is possible to work backward and calculate when they might all have been squeezed into a single spot. This is the basis of the "big bang" theory of the origin of the universe, and the best current estimate suggests that the expansion, whatever caused it, began about fifteen billion years ago. That is an unimaginably long time, but to put it into somewhat more personal perspective, we now know from radioactive dating of the oldest rocks on our planet that earth itself is about five billion years old. Call that length of time something simple, like one "geo," and our galaxy turns out to be about two and a half geos, and the cosmos a little more than three geos old, which is much more reassuring. We may not have been in at the very beginning, but neither need we feel like total upstarts.

Steven Weinberg, an astrophysicist at Harvard, has put most of our current knowledge about the origins of the cosmos together into a fascinating scenario for "The First Three Minutes" of the big bang.[575]

It is clear that we can't think about it as a sort of gunpowder blast, which starts from a definite center and spreads out to engulf things around it. It was an explosion "which occurred simultaneously everywhere, filling all space from the beginning, with every particle of matter rushing apart from every other particle." In fact, in the beginning, during the first one-hundredth of a second, the temperature of the universe was about a hundred billion degrees Centigrade, which was too hot for the particles to have any stable existence at all. They were annihilated as fast as they were formed.

This storm of creation and destruction continued to rage in the cosmic teacup for about three minutes, until the temperature dropped to a mere one billion degrees, which was cool enough for nuclei of hydrogen and helium to begin to form. And only later, much later, about seven hundred thousand years after the bang, were these nuclei able to join with stray electrons to produce the first actual atoms of gas. Then these, under the influence of gravity, formed clouds or clumps that ultimately condensed into galaxies.

This is our current myth, and the best available evidence for it, for the fact that there was a big bang, is that we can still hear the echoes. The cosmos is bathed in background noise. There is a constant sort of dismal universal hiss, like static on a radio, that seems to be the distant reverberation of that early explosion. The cosmos preserves a memory of its origins so faint that it wasn't even noticed until 1965. And only in the last few months has it been accurately measured by means of an ultrasensitive microwave receiver in a U2 spy-plane flying steady at sixty-five thousand feet, above atmospheric interference.[229]

The most interesting feature of this background radiation is that it is extraordinarily uniform. There is a slight decrease in the direction of the constellation Aquarius and an equivalent increase in the opposite area of Leo, which tells us that our galaxy is traveling Leoward at 1.3 million miles an hour. But the overall levels are otherwise the same everywhere. The big bang, that most cataclysmic of all events, now seems to have been a strangely gentle explosion, not so much a violent blow-out as a finely orchestrated event—more like the steady inflation of a football than the burst of a bomb. It begins to look as though the cosmos might just be breathing out, and when the current exhalation ends, perhaps after another ten geos, it will pause and then inhale. The process will begin to run backward, giving way to an accelerating contraction that can only end when everything approaches a new state of infinite density; when there will be another big bang and another grand expansion.

If that is our future, then it must also be our past. Everything seems to point not only to a cosmos that evolves, that grows and changes, but perhaps to an oscillating breathing system, one that keeps on bouncing back. And if we had the vision to see the cycles of expansion and contraction in a larger universal perspective, it seems likely that this too would prove to be an evolutionary one. It is hard for the flea on a dog to distinguish one breath of its host from another, harder for the flea to conceive of the part each inhalation plays in the total lifespan of the dog, and inconceivable that the insect should even begin to appreciate that this individual mammal is but one of millions playing an ongoing role in the evolution of a species.

The spiral soars into incomprehension however one looks at it, but the concept of an organic system, of a cosmos that lives and breathes, is I think a useful one. It helps us to think of life on earth as part of a larger pattern, instead of as something freakish and unique that could only happen here. It helps too to shake an assumption implicit in much thinking about the origins of life—namely that organic evolution must necessarily take place on a planet like ours in association with a star. The latest evidence not only is heavily in favor of extraterrestrial genesis, but also avoids any connection with heavenly bodies by siting the cosmic Garden of Eden in the void, somewhere out in space between the stars.

We owe most of our knowledge about the fine structure of the cosmos to a German optician who was the first to notice that the spectrum of colors produced by refraction of sunlight through a prism was punctuated by solid dark transverse lines. Joseph von Fraunhofer in 1814 detected and mapped the position of several hundred such lines, which are still named after him, but it was left to the physicist Gustav Kirchoff in 1859 to show that each pattern of lines was peculiar to a particular element. Today spectroscopy recognizes more than ten thousand lines and can detect the distant presence of almost any substance by its remote but characteristic fingerprint. A simple optical technique has taught us as much about the cosmos as any of the greatest telescopes.

We have learned, for instance, that space is far from empty. The gaps between the stars are punctuated by clouds composed largely of gas and dust, which occur in the ratio of about a hundred to one.[471] The cosmic dust grains themselves are about the same size as the wavelength of visible light, and, where they are numerous, they appear as dark or luminous nebulae, usually in the spiral arms of galaxies. The soaring Horsehead nebula in Orion, rising like a rampant equine cloud, is a dramatic example of opacity produced in this way. Elsewhere the grains are sparse and almost undetectable.

If the dust has the same composition as the interstellar gas—and there is no reason to assume otherwise—it is largely made of molecules of carbon ($C$), nitrogen ($N$), hydrogen ($H$), and oxygen (O). This provides an interesting, and perhaps meaningful, coincidence. The grains in the interstellar medium are the same size and have the same composition as some of the most simple living things on earth, our terrestrial bacteria. Even the most complex living organisms, such as human beings, are not in essence very much different. The same four elements provide ninety-six percent of our body weight, much of it in solution. As J. B. S. Haldane observed, "Even the Archbishop of Canterbury is sixty-five percent water."[464]

Naturally it is not enough just to have the right elements in the right proportions. Mud can do that. Something extra has to be added to breathe life into dust. New arrangements are necessary, and because carbon is one of the four basic substances, a multitude of patterns is possible.

Atoms of carbon each prefer to be linked up with four other atoms, either of their own or other kinds. And they can do so with single or double bonds, in clumps, lines, chains, branched chains, or even complete rings, to produce molecules with anything up to several million parts. These are the stuff of organic chemistry, the study of compounds producing, or produced by, living organisms; and these are the kinds of reaction which it now seems certain are taking place between the stars.

Awareness of this possibility seems to have reached a peak at a gathering of astronomers in Cambridge, Massachusetts, in 1973.[217] Several papers presented to what is now known as the Dusty Universe symposium pointed out that there seemed to be a lot of atoms missing from interstellar space. Spectroscopic analyses kept coming up with results that showed less carbon, oxygen, and nitrogen than everyone expected to find there.

New models were put forward, and the most promising of these suggested that the missing atoms had gone undetected because they were bound up on the surface of interstellar dust grains in a sort of molecular mush which Mayo Greenberg called "dirty ice."[218] At that time there was little experimental evidence for the existence of such accretions on cosmic grains, but since then, radio astronomy—which receives and interprets microwaves rather than light waves, and can look right through interstellar clouds—has given us what we need. The very short radio waves provide a sort of electronic spectrum that includes fingerprints as distinctive as those that appear in the lines of an optical spectrum. And in the last few years radio astronomers have detected an ever-increasing list of simple organic molecules in interstellar space.[254]

The first substances detected in this way were nothing more

than simple connections of the most common atoms: cyanogen (CN), carbon monoxide (CO), and hydrogen cyanide (HCN).[579] But then formaldehyde ($H_2CO$) was found, and before long formic acid (HCOOH), methanol ($CH_3OH$), acetaldehyde ($HCOCH_3$), and methyl formate ($HCOOCH_3$) turned up. One doesn't have to be a chemist to see that the progression is toward ever more complex organic compounds.[470]

It has also very recently been suggested that some mysterious features observed in parts of the spectrum could be accounted for by assuming that, in addition to these mixed organic molecules, long chains of pure carbon are also accumulating to form what amounts to a coating of sticky tar on the surface of the cosmic dust.[133] This would make it possible for dust grains which bump into each other to adhere and form grain clumps in which even more complex chemistry was possible.[266] And an additional advantage of interstellar tar is that, like the tarmac on a highway on a hot summer day, it absorbs ultraviolet radiation and raises its temperature above that of the environment, which in free space approaches absolute zero ($-273°$ C).[317]

At the surface of a cosmic dust cloud, a little heat and a lot of radiation could produce reactions in which molecules disintegrate and recombine in all sorts of new ways; but an even more important process could also take place in the cold heart of dense interstellar clouds.[207] It used to be assumed that chemical reactions need heat and slow down as temperatures were reduced. This is true, but only up to a point. We now know that as the temperature nears absolute zero, a strange thermodynamic inversion takes place and many processes actually accelerate, so that the complex early evolution of carbon compounds is more likely to take place inside interstellar clouds than almost anywhere else.

The most widely accepted model of star formation assumes that interstellar clouds gradually condense as the dust grains fall together under the influence of gravity.[96] Clumps of grains will tend to fall more quickly, and the largest ones will soon accumulate in a layer at the midplane of the cloud, which now becomes a primitive solar nebula. Further instability at the core then forces the clumps together into even larger bodies, the size of asteroids, and as these in turn aggregate at the center, a protostar is formed. This continues to contract, all the time getting hotter and hotter, until a density and temperature are reached where thermonuclear reactions begin and the mass becomes a star and starts to shine.

Some of the grain clumps, together with their growing organic factories, inevitably get cooked, but during the early stages in the birth of a star, the outer cloud remains comparatively cool and its carbon molecules should be unaffected. It is these layers with their attendant molecules that form the protoplanetary mass which in turn condenses and coalesces to give rise to planetismals and, in time, proper planets. The new planetary area remains rich in dust and complex organics, and as the central star warms the system, normal chemical reactions speed up and a steady stream of carbonaceous material showers down on any orbiting body.

According to this model, a favorable planet with a water-rich surface would have the potential for evolving life at a much earlier stage than we usually imagine. This may have been how our solar system came into being, but there is still no general agreement on that count. There is evidence that our sun could be a second- or even third-generation star, drawing its raw materials from the scattered remains of distant super-novae—in which case our planets could have a totally independent origin, perhaps even in the remains of a double star, twin brother to our sun, which blew up at an early stage in their fraternity.

One day we will solve that particular riddle, but at the moment and for the sake of this discussion, it doesn't really matter how the earth was formed. What is important is the knowledge that wherever the planet came from, it is still being showered with the products of interstellar condensation.

Every day the earth collides with more than a hundred million meteors and something like a hundred tons of extraterrestrial material comes pouring into our atmosphere.[354] Most of the intruders burn themselves out as shooting stars in friction with our air, but a large part of the crust of the planet is nevertheless made of cosmic silt, about a ton each day settling slowly on the surface. Anything which reaches the ground is called a meteorite. Most are microscopic, but some are quite substantial; the biggest ever recovered intact was found in Greenland and weighed thirty-seven tons. Approximately half of all the macroscopic meteorites are metallic, mainly iron and nickel, and the other half are composed largely of silica or stone.

Most of the stony meteorites contain small glassy inclusions, and about two percent of these are called carbonaceous chondrites because they contain significant quantities of organic matter.[8] The proportions, in fact, are extraordinarily high. About 0.1 percent of all material which has ever fallen on earth is organic. By comparison, if we measure the total weight of all organic matter on earth against the mass of the planet itself, only 0.0000001 percent is of living origin. This means that meteors are coming from somewhere that is a million times more organic than earth itself—which is something one has to stop and think about for a while.

Several rigorous analyses of carbonaceous chondrites have now been made, and all show quite clearly that they contain compounds such as paraffins, long-chain aromatic hydrocarbons like tar, fatty acids, amino acids (the basic precursors of protein), and even porphyrins (the building blocks of chlorophyll).[265] And in early 1977, an international group of seven scientists—astronomers, chemists, and applied mathematicians—collaborated in an intensive study on a recently collected chondrite from a site in Africa.[472] They found that it contained an organic compound, an aromatic polymer, whose spectral properties are identical to those long known from interstellar extinction curves. The fingerprints are unique and unmistakable. For the first time we have proof that some meteorites have their origin in presolar interstellar clouds and must be among the most primitive solid bodies in the universe. And, more than that, we have evidence of a cosmic trade in which complex organic compounds, precisely those necessary for the initiation of life, are manufactured in space and imported here to earth.

If you listen carefully, you can almost hear the seed being sown.

I choose that analogy deliberately because it is possible that the goods we get in this way may be more than mere fertilizer; they may be gift-wrapped packages, sort of cosmic cocoons, containing the seeds of life itself.

When the world got over the shock produced by Darwin and was forced to abandon the notion that creation occurred one afternoon in 4004 B.C., scientists began casting around for a new paradigm. Perhaps because he lived in an officially atheistic society, predisposed to the acceptance of mechanistic notions, Alexander Oparin of Moscow University was the first to come up with a comprehensive chemical theory to account for the development of life on earth from inanimate matter.[410] He suggested that the primitive earth surface was subjected to constant energetic bombardment by electrical storms, which sparked connections between simple inorganic molecules and knitted them into larger, more complex prebiotic substances. And that these new molecules survived because the early atmosphere was without oxygen, which, in normal circumstances, would tend to break them down.

As rust and fire show quite clearly, oxygen is a highly reactive gas. It combines rapidly with other molecules to form new chemical compounds, and in the inorganic world this reaction is inexorable and irreversible, a one-way street as long as the oxygen lasts. When a material combines with oxygen, it is said to be oxidized. Materials with a high hydrogen content, on the other hand, are said to be reduced. Organic matter is one of the latter and therefore all living things are, by definition, the end products of reduction. Yet we live in an atmosphere of oxygen. In a very real sense, we on earth are living in a poison gas. Worse still, some of us actually breathe the stuff.

Despite its toxicity, oxygen has useful properties, and most complex organisms have evolved very sophisticated ways of protecting themselves from its corrosive effects. We generally succeed in avoiding direct contact with it altogether, delegating this responsibility to specialist employees who make the gas work for us by breaking down food in a process of respiration. We have even come to rely on their services to such an extent that we can no longer survive without oxygen. We, and most living things on earth today, are obligate aerobes. But it seems likely that our elaborate specializations are comparatively recent developments and that the first living things probably had no such defenses.

There still are organisms which are poisoned by oxygen. They live in the few places, such as deep soil and oceanic mud, and the interior of some parts of other organisms, where oxygen is almost absent. None of them are more advanced than simple worms, and we group them all together as obligate anaerobes. It was another Marxist, though this time not an obligatory one, who suggested that these could be direct descendants of the first living things, relics of a time when conditions were anaerobic.[231] J. B. S. Haldane felt the synthesis of hydrogen-rich organic compounds would be much easier to understand if their environment had been a reducing one, without oxygen. So today this theory of an early world which had to wait for the advent of plants to provide oxygen is known as the Oparin-Haldane model, and it has been the jumping-off point for most of our experimental attempts to produce new life in the laboratory.

Haldane saw the cradle of life as a primeval soup, a sort of hot dilute brew of simple molecules.[51] So in 1953 at the University of Chicago, Stanley Miller took sterilized water, added an atmosphere of hydrogen, ammonia, and methane, and subjected the mixture to continuous electrical discharge for a week. The results are history.[377] He got several simple amino acids, and by inference, if this was possible in one week in a single simple setup, then it was likely that a billion years in the primordial ocean would be time enough to give rise to more complex nucleic acids, even those capable of continuous self-replication.[378] In the last twenty years, there have been many more sophisticated experiments.[183] Ribose and glucose sugars,[450] long chains of amino acids,[238] and even nucleoside phosphates[435]—the precursors of DNA—have all been manufactured by simulating what are believed to have been the primitive earth conditions. But these are simulations, all under carefully controlled conditions, most ending precisely when the experimenters get what they want. There were no such controls or restraint in the early years of our planet, and these experiments do not necessarily constitute proof of the Oparin-Haldane theory. There is no astronomical or geological evidence for an early reducing atmosphere and no trace of carbonaceous deposits in the oldest sedimentary rocks. Primitive earth conditions may not have borne any similarity whatsoever to those used in the laboratory. There may never have been a primeval soup.

Fred Hoyle and Chandra Wickramasinghe of University College, Cardiff, believe that life was, and still is being, brought here from an extraterrestrial source.[267] They regard prestellar molecular clouds, such as those in the Orion nebula, as the most natural cosmic cradles and believe that "processes occurring in such clouds lead to the commencement and dispersal of biological activity in the Galaxy." And they argue that interstellar clumps of dust grains bearing polymers become involved in competitive situations which lead, in a Darwinian sense, to evolutionary developments and to the survival only of those that are fittest. "The simplest self-replicable system involving clumps of inorganic grains glued together with organic polymer coatings becomes most widespread in the Galaxy," with the result that, given the right conditions, "the organic polymer films, which separate individual grains as well as surround entire clumps, evolve into biological cell walls." And finally these systems spit out their dust grains and become primitive and interstellar cells.

This is a very novel and radical idea which still involves a great deal that is purely speculative. In essence it bears a startling resemblance to a successful science-fiction story published more than a decade ago by Hoyle himself which tells of a black cloud that envelops the sun and turns out to be intelligent; but yesterday's science fiction has a way of becoming today's established scientific fact. The new experimental data on the connection between meteoritic and interstellar spectra does seem to lend respectability, even a degree of credibility, to such speculations about the origin of life on earth.[594]

In 1961, Bartholomew Nagy and George Claus of New York University caused a furor when they published a paper claiming to have found microfossils in a meteorite. They described small circular forms, shield shapes, cylinders and hexagons that could be seen in sections of two separate carbonaceous chondrites that fell in 1864 and 1938. They said, "The fact that the same types of organized elements were present in both Orgueil, which fell in the temperate climate of Southern France, and in Ivuma, which fell seventy-four years later in an arid tropical region of Central Africa, renders unlikely contamination with morphologically identical, locally derived, microorganisms."[112]

Everyone with access to suitable meteorites and high-powered microscopes began to look for similar evidence. Some were unsuccessful and criticized Nagy and Claus on the grounds that most of the shapes they described could equally easily have been produced by artifacts arising in the cutting, grinding, staining, and mounting of their specimens,[7] though even they admitted that there were some simple shapes "definitely indigenous to the meteorite" which had no known terrestrial equivalent and belonged "in a morphological no-man's land."

One of the most serious barriers to proof that the inclusions really do have an extraterrestrial source is the fact that many of the meteorites are porous and probably "breathe" on their way down through our atmosphere. Two researchers at the University of Chicago showed that ordinary ragweed pollen, when prepared and stained with the same procedure used by Claus and Nagy, produced shapes very much like some of their "organized elements." Faced with a choice between assumptions—either ragweed flourishes on asteroids as in *Le Petit Prince*, or the Orgueil and Ivuma meteorites were contaminated by pollen on arrival here—they not unnaturally chose the latter.[487] It isn't a problem that can easily be solved.

A more recent study on the insoluble parts of the famous Orgueil meteorite (there can't be much of it left now) shows that it contains a chemical called sporopollenin which has also been found in a microfossil planktonic algal spore from a Tasmanian rock three hundred and fifty million years old.[80]

The fossil has been put into a terrestrial scheme of things by giving it an orthodox scientific name—it is called *Tasmanites punctatus*—but Chandra Wickramasinghe has grander plans for it. He says, "It is possible that such spores represent primitive interstellar protocells in a state of suspended animation. Their release into the Earth's atmosphere may have led to the start of all life on our planet."[588]

He could be right. Ten years ago two Harvard geologists discovered a minute, bacterialike rod-shaped organism in sedimentary black chert exposed by a road cutting near Barberton in South Africa.[34] Nothing odd about that; geopetrologists keep finding simple forms of life in their survey samples—some species are even considered to be indicators which signal the oil-bearing potential of deposits. But the extraordinary thing about the Barberton find is that it was made in a Precambrian rock that is more than three billion years old.

There are two ways of taking that. Either this is the oldest known evidence of biological organization in the fossil record; or else *Eobacterium isolatum*, which as its name suggests is embarrassingly isolated at a very early stage in the history of the planet, comes from somewhere else. It is certainly a true living organism—nobody quibbles about the status of bacteria as proper self-replicating life forms—but it is equally certainly very much like the cylindrical organized elements found in their meteorites by Claus and Nagy.

At least one further source of information about the strange things in meteorites remains to be explored. If the organic compounds are protocells in a state of suspended animation, perhaps they can be roused. Soviet and American scientists have been trying to do just that.

Fred Sisler of the United States Geological Survey has begun collecting samples from the interior of carbonaceous chondrites, and he finds that even after a long period under sterile conditions, some of his nutrient broths nevertheless cloud over, indicating the presence of living microorganisms. And at least one of these sleeping beauties, roused from an unimaginable slumber, is totally unfamiliar to terrestrial microbiologists. No one has ever seen anything like it here before, so it is going to be hard to dismiss that one as a contaminant.[487]

One of the difficulties facing the earliest life forms, whether they originated here or elsewhere, would have been their rarity. A single cell or protocell couldn't have lasted very long, even if it was an efficient replicator. It would have been swallowed up and lost forever in anything as large as a primordial ocean. The whole chain of life did not derive from a single Adam-cell, blessed with the vital essence. We must have had multiple ancestors. Of course, we don't know how often life began and was snuffed out again almost before it got started, but it seems certain that when it did eventually take root here, it managed to do so by crossing some sort of density threshold —by having enough of it around to be able to cope with a certain amount of attrition.

The random arrival of isolated bits of life in protocells that manged to survive both the long icy journey and the fiery welcome probably wouldn't have been sufficient—not unless a whole mass of protocells managed to arrive simultaneously; and as far as we know, there is only one way that could have been organized: if they hitched a ride on a comet.

"Comets are the nearest thing to nothing that anything can be and still be something."[11] They are "all cry and no wool . . . mere luminous vacua."[490] In fact, they consist of parcels of cosmic dust and gas, ammonia, methane, and water ices all arranged into a nucleus like an orbiting snowbank a few miles across to which is appended a diffuse tail perhaps a million miles long. As far as we can tell, they breed in interstellar space and only flash across our skies when perturbed by passing stars which send them into orbit around our sun. Some, on a hyperbolic orbit, visit only once, like celestial outlaws; but others have an elliptical orbit that brings them back again and again. It was Edmund Halley, professor of geometry at Oxford and a friend of Newton, who noticed the pattern and predicted that a comet he had personally seen in 1682 would return again in 1757. It did, fifteen years after his death. And right now Halley's comet has turned again at aphelion a billion miles out in the void and is rushing back like a great wild satellite of the sun. We will see it in 1986.

It seems that comets have their origin in clouds of helium and hydrogen which are expelled from our sun, and from other stars, in their early overluminous enthusiasm. These vapors, along with a fine smoke of ices and silicates, drift through the cosmos and, in their passage, mop up masses of interstellar material like giant sponges. Gradually this material coalesces into a distinct though insubstantial core and prebiotic molecules known to exist among the dust grains condense around nuclei of ice to form the core or coma of a distinct comet. It is possible that organic molecules may make up as much as thirty percent of the mass of a comet—a concentration greater than that known for any other situation in the universe, including the surface of the earth.

It has often been thought that our planet owes its atmosphere to a direct collision with a comet in which we, the larger body, captured and kept the materials of the smaller one under the influence of our gravity. Now Fred Hoyle suggests that there have been other close encounters and that terrestrial life "could well have originated about four billion years ago by the soft landing of an icy comet already containing primitive organisms."[267]

It is a big jump from prebiotic molecules in interstellar clouds to primitive organisms on a comet, but it is not an unreasonable one. When a comet gets anywhere near the sun, its water melts and could mingle with the trapped dust to produce a solution of organic molecules which, we know from spectroscopic analysis of Kohoutek's comet in 1973, includes amino acids and heterocyclic compounds. Then the comet moves away from perihelion again and cools. This oscillation in temperature, a periodic cycle of melting, evaporation, and refreezing, provides one of the most vital ingredients for any evolutionary system—a fluctuating environment that can act as a selective pressure to encourage adaptation and change. On a cometary nucleus, selection will favor those molecules best able to withstand breakup during the hottest phases, and freezing during the alternating periods of intense cold. During the passage past the sun, ultraviolet radiation on the melted surfaces of the comet would tend to produce polymerization reactions, leading to the elaboration of more complex macromolecules. And all such new inventions produced in one cycle would be remembered; they would be held in a frozen state as prescriptions for survival until they were needed again during the next approach to the sun. They would in effect be like genes, particles with memory, helping to develop simpler molecules into the same predetermined structures best suited to long-term survival. There is no reason why a process such as this should not go on, cycle after cycle, for millions of years. There is every reason to presume that, out of such an evolutionary system, life could well emerge.

Hoyle believes it did and, perhaps a little optimistically, suggests that even bacteria could result. He sees a photosynthetic bacterium as one possibility, growing near aphelion when water was still warm enough to be liquid but sunlight was scarce, and at the other extreme, with the comet near perihelion, perhaps a thermophilic bacterium like those found now in hot springs.

If things did indeed happen in this way, there would be few difficulties involved in transferring life forms from the comet to a suitable substrate on a planet such as ours. With every passage near the sun, all comets are stripped of their surface layer, parts of which volatilize and are blown away by the solar wind, but much of which spreads out in the long diffuse tail and inevitably adheres to any solid body, like the earth, which passes through this enormous vapor trail.

This concept of comets as cosmic storks is revolutionary as far as science is concerned, but it is not new to older folklores. Among the Navajo of North America, comets are the spittle of Black God, creator of the stars, and carry his seed to the dark parts of the sky.[230] The Bushmen of the Kalahari say that comets are hot ash of a scented root thrown up by the Girl of the Early Race as she goes about her business of making the stars.[60] In the Amazon, the Tukano believe that shooting stars and comets are signs of celestial sex and that from their union, fertilizer falls to the earth as dew.[449]

In all times and places, comets have been regarded with awe and often with some trepidation. Many people both ancient and modern see them not only as sowers, but as bearers of evil seed, harbingers of pestilence and destruction. It is not surprising that anything as untoward, as apparently unpredictable, as a comet should be looked at askance. And it is not unexpected that these misgivings should be associated with illness and death. Even science has done little to dispel such feelings. When Halley's comet last appeared in 1910, and it was announced that we would pass through the invisible tail, which was composed of poison gases, it was widely believed that everyone on earth would asphyxiate. Thousands of people were actually left feeling distinctly ill after the comet disappeared, but this probably had a great deal to do with a series of massive sybaritic end-of-the-world parties. Or did it? Can we be certain that there is no truth in the traditional connection between comets and disease?

It is generally believed that major pandemics, such as worldwide outbreaks of new forms of influenza, begin with the random mutation of a virus at one place, which then spreads by direct person-to-person contact. This is a convenient model which gives health authorities something to work on, but it is by no means well proved. The earliest identifiable record of a "flu" is one from England in 1173. Between that date and 1875 there were ninety-four epidemics, and of these at least fifteen were of pandemic proportions, affecting most of Asia as well as Europe, despite the fact that communication and the movement of people was slow and in many areas virtually nonexistent.[257] Similar descriptions of sudden onset and rapid global spread seem to be characteristic of many early as well as later epidemics. The reported speed of transmission of the diseases is difficult to understand if, as is usually supposed, infection passes only from person to person or through vectors such as rats and lice.[600] Air travel today complicates the picture considerably, but prior to the age of internal combustion, it is almost impossible to account for rapid pandemic growth without assuming the existence of some other kind of agency.

Fred Hoyle and Chandra Wickramasinghe believe they have the answer. Suppose, they suggest, extraterrestrial biological invasions are a regular occurrence. "These invasions could take the form of new viral and bacterial infections that strike our planet at irregular intervals, drifting down onto the surface in the form of clumps of meteoric material." These then produce epidemic diseases, each of which represents "renewed attempts at the evolution of life on comets, infection reaching the earth when its orbit crosses the trails of debris from these comets."[267]

Reports of the sudden spread of plague and pestilence punctuate the history of all countries, but it is important at the outset to know just what is meant by a plague before deciding that its origin could be extraterrestrial.

Man is host to a wonderful zoo of creatures of all shapes and sizes. "Microbes cling to us where they can, living in their own way, if they are up to it, off the land and the windfalls. There they grow, usually with a certain moderation, varying with the local climate and especially with the relative abundance of the food supply."[465] The emphasis for all successful symbionts and parasites is one of moderation and good ecology. Killing your host is bad for business. And conversely, a human body that resists infection so completely that the would-be parasite cannot find a toehold anywhere is creating another kind of crisis for the infectious organism. That sort of pressure could push it into an adaptation which would be far more detrimental to the host. So, in the millions of years that they have had to get to know each other, what usually happens is that parasite and host reach some kind of reasonable accommodation. They compromise to their mutual advantage.

All of which means that we can assume, right from the start, that a lethal disease-producing organism is poorly adjusted. It is still in an early stage of adaptation to its human host—which in turn implies that it must be new to the job and could, once we have exhausted several other possibilities, even be new to the planet.

Plague in medieval times meant *Pasteurella pestis*, a bubonic bacillus that appeared very suddenly and brought the black death to millions, effectively halving the population of Europe and changing history. It was carried by fleas which spread on the backs of black Indian rats that traveled wherever ships could take them, but it was no better adapted to the rats than it was to human hosts. It killed them both indiscriminately, and after four centuries of destruction, vanished almost as suddenly as it had appeared.[399] On the surface this looks a likely candidate for cometary origins, but careful detective work by modern medical geographers has shown that the bacterium is an old earth resident and still exists in the burrows of wild rodents in Central Africa, to whom it is perfectly adapted. For these hosts, a bubonic infection is no more serious than a childhood disease like mumps or measles is for us. They recover from it and are left with a lifelong immunity. It seems that the bacterium became a plague not because it or the human host was new to the earth, but because it found itself in a new situation. The environment changed when people took not only to traveling more widely, but also to returning to homes in poorly designed, overcrowded, rat-infested cities where cross-infection was almost inevitable.

The plagues of smallpox which ravaged Mexico and Peru, enabling Cortez and Pizarro with mere handfuls of men to conquer the Aztec and Inca empires, must have seemed like supernatural intervention to the Indians, who saw only that they died while the Spaniards were left unharmed. We now know something about acquired immunity and can see that this too was a good example of an old pathogen busy exploiting new pastures. No cosmic solution need be invoked; but we are on very much more difficult ground when it comes to explaining the rise and fall of illnesses that have no intermediate hosts, seem to need no change in ecology, and provide no personal immunity.

The biggest problem is the most obvious one—that range of respiratory disturbances that, depending on severity, we label influenza or just a common cold.

Practically, all that can be said with any certainty about the cold is that it is common, standing almost permanently at pandemic level, and that it is almost impossible to describe and classify. At least thirty viruses are believed to be involved, and vaccines have been prepared to provide short-term immunity against some of them. The problem, however, is that they either alter details of their chemistry at frequent intervals, in what looks like a frantic attempt to adapt to and settle down in a new environment, or they are constantly replaced in the front lines of the battle by totally new forms of virus that suddenly appear. We in the West like to think that each new wave of sneezing and coughing has its origins in some distant, probably less hygienic place like Hong Kong; but in Asia they ascribe the same symptoms to "English" or "American" flu. The truth is that nobody knows exactly where the new forms come from. They could quite easily be extraterrestrial.

Perhaps the best evidence for the invasion of infection is the fact that major pandemics tend to be short-lived, usually lasting not much more than a year, and that they never end up by affecting the entire population. Theoretically, each time a new virus reaches a new community it should wreak the same havoc as it did in the first area to be infected. Old established parasites like smallpox do just that, but completely new diseases seem to lose their potency as they get filtered through a certain number of human hosts. Hoyle and Wickramasinghe contend that "primary cometary dust infection is the most lethal, and that person-to-person transmissions have a progressively reduced virulence, so resulting in a declining incidence of the disease over a limited period."[267]

They go on to suggest that "the abrupt appearance in the literature of references to particular diseases is also significant in that they probably indicate times of specific invasions." It is interesting too that many of the earlier plagues, such as the one which devastated Athens in 429 B.C., and another that killed two out of every three people in central China in 312 A.D., seem as far as we can tell from contemporary descriptions to have no modern counterparts. It is possible that rare diseases, outbreaks of which are separated by long periods of time, could appear whenever our planet passes near a long-period comet, whereas minor variations of a common infection, such as influenza, could be due to more frequent, regular passages of the earth through the track of shorter-period comets.

It is difficult to draw connections between specific comets and definite diseases. There seems, for instance, to have been no direct measurable effect following each of the calls of Halley's regular visitor, unless one recognizes hysteria as a characteristic symptom. But there is no reason to assume that we need actually see the guilty party. Epidemics could result when the earth passes through debris left by comets now long gone, or they could be seeded by comets and take a considerable time to ripen.

The factors which control the actual pattern of infection produced by any particular extraterrestrial invasion could also be very complex. If the infectious organisms are dispersed in a diffuse cloud of small cometary particles, the onset of the disease may well be rapid and its incidence global. On the other hand, an aggregate of infective grains in clumps falling over a limited land area would tend to produce a more localized effect. It is likely that air currents and prevailing winds play an important part in controlling dispersal and that certain latitudes might well be more directly affected than others. It could be significant that outbreaks of influenza tend to be particularly heavy in areas associated with the "roaring forties" —latitudes that are fed by permanent strong airstreams and that cover the United States, most of Europe, Korea, Japan, and as remote an area as New Zealand. Influenza certainly is very common in all these areas, but that could also have something to do with the fact that they tend to be densely populated.

Furthermore, local atmospheric conditions such as humidity or temperature inversion, or even air pollution, could produce marked variation in the time the inefective dust took to settle. And this could produce a pattern which appears to focus the blame for the spread of the disease on one particular terrestrial center, such as much-maligned Hong Kong, which in fact had nothing to do with the development of the new virus at all.

This speculative connection between comets and disease will need to be supported by a great deal more hard evidence before it is taken seriously by epidemiologists. All that can be said about it with any certainty at the moment is that it provides an intriguing model which reinforces some of our oldest and most strongly held folk beliefs. It also provides a possible mechanism for the distribution of organic seeds, generated at one point in the galaxy, to all other points, and stresses the possibility that very simple living things, or at least their direct precursors, could have arrived here early in our history to spark the evolutionary process, and could still be arriving here in many ways that produce local, or even lasting, effects on our developing biosphere.

I believe it is also important for a more personal human reason: Taken together with what we are beginning to learn about the origin of the universe, it serves as a forcible and timely reminder that our planet may be unique in many ways, but that we do not live in a sealed spacecraft, isolated from the environment in our convenient bubble of air. Earth travels rapidly through space and time, constantly exposed to the complex ecology of our galaxy, which includes comets and interstellar debris. And the space between the stars is punctuated by molecular clouds which contain organic compounds that could provide all that is necessary for the generation of self-replicating life.

It emphasizes the fact that the cosmos as a whole is subject to universal laws and influences and shows a remarkable uniformity, more like an enormous rhythmic organism than a disparate collection of unrelated fragments hurtling away from each other.

Looking at ourselves in this way, we may begin to feel less special, but we should also feel a great deal less lonely. "Man," said Henry Thoreau, "is but the place where I stand, and the prospect hence is infinite."

Life on earth begins to seem less like "an offensive directed against the repetitious mechanism of the Universe"[585] and much more like an integral part of a vast ongoing evolutionary process. A seed thrown up, and taking vigorous root, in the soil of a friendly shore.

# CHAPTER TWO

# THE SOIL: Life on Earth

Dust to dust. The imagery is vivid and the decay inevitable, but it is difficult to imagine man arising from the dust of the ground. Unless you believe in miracles. Or unless you happen to be a chemist at the University of Glasgow.

The greatest barrier that life, or any theory about its origins, has to overcome is the development of a replicating system. It is clear that simple organic compounds can and do originate spontaneously from basic elements available almost anywhere in the universe.[268] And it seems certain that some of these do, by chance and under the influence of cosmic radiation, combine to form even fairly complex molecules such as amino acids. But it is an enormous jump from there to the existence of anything like a protein, and the barrier is equally formidable whether you face it here or on the nucleus of a comet.

To get some idea of the inherent improbability of even a relatively simple protein, consider insulin, which consists of twenty different amino acids arranged in a particular configuration fifty units long. Imagine a planet like earth without life, but with the twenty vital acids growing up, or raining down, on its surface. Assume that solar radiation exerts sufficient force to cause these acids to arrange themselves into random chains, a new one being formed about every tenth of a second. And let this go on uninterruptedly for five billion years. What do you think you'll get?

Alexander Cairns-Smith at Glasgow calculates that "even if the whole earth had been made of nothing but amino acids which had rearranged themselves randomly and completely ten times a second for the whole period of the earth's history, there would have been little chance of producing even once, for a tenth of a second, one molecule of insulin."[91] A protein is a structure of such gigantic improbability that unguided nature would probably not hit on it even if given the whole known universe to experiment on for a billion years. The odds against it happening by sheer chance are greater than one in $10^{80}$, which is a figure larger than the total number of electrons in the universe.[145]

So obviously blind chance received a helping hand; but whose?

The classic scientific theory of the origin of life, the Oparin-Haldane model, assumes that a long period of chemical development took place prior to the beginnings of organic evolution. Oparin says that "matter never remains at rest, it is constantly moving and developing and in this development it changes over from one form of motion to another and yet another, each more complicated and harmonious than the last. Life thus appears as a particular very complicated form of the motion of matter, arising as a new property at a definite stage in the general development of matter."[410] This is an appealing view of a process with an inherent tendency to drift toward an organic goal, but it doesn't explain how a random system can suddenly turn into one capable of replicating itself. It is not clear that an evolutionary process without replication must inevitably, or can indeed ever, lead to one that does include it.

It is possible, as Hoyle suggests for comets,[267] and Oparin for the early earth,[410] that an evolving system could provide a particularly suitable environment within which some kind of simple self-replicating control device, perhaps a sort of protovirus, could arise. But it is equally likely that the same system, simply because it is evolving and changing, would succeed only in destroying all new and useful combinations as fast as it gave rise to them. Stockpiles would tend to decompose rather than accumulate. In the presence of high radiation, the longer the early biochemicals lay around, the greater the chance that they would simply have deteriorated into hopeless "gunks" like raw petroleum—which may be precisely how our oil deposits arose.

Cairns-Smith has great reservations about the possibility of life arising in a tarry soup and suggests that early metabolic processes "would have been seriously hampered by a richly reactive chemical environment." He feels that any sugar that happened to be formed would be equally quickly turned to sticky caramel. And he suggests that "there need never have been a pre-vital soup or, if there was, no historical connection between it and us."[93]

His most serious objection to the classical theory is that it assumes, without good reason, that the central biochemical machinery which is common to all life now was always there, and arose in the beginning as a result of some sort of physico-chemical accident.

It is easy to understand how most theorists came to adopt

the old model; it is impossible not to be impressed by the uniformity of metabolism in all living things. Life produces a mind-boggling multitude of special effects, but perhaps the most extraordinary thing we have yet learned about it is that it does so with a surprising small bag of tricks. Everything boils down in the end to the presence in all living things of four substances called nucleotides, which are knitted together into the elegant double-helix shape of a molecule of deoxyribonucleic acid—the famous DNA.

This is the common factor. DNA occurs only in living things, and it occurs in every one of them. There are differences in detail in the way in which the four nucleotides are strung together, and it is these subtle alterations that produce the variety of living organisms; but the bases are identical whether they occur in an apple or an ape, in a mouse or in a man. There is a basic unity of all metabolism, and it is highly significant that DNA, the unit itself, is the molecule responsible for the control of replication.

Replication provides continuity, but that is not all it does. If it were, the world would be filled with a large number of identical replicas of one original organism. But things aren't like that, simply because no copying system can be perfect; because mistakes will always happen—must happen, because it is these mistakes themselves that make evolution possible.

Each small mistake produces a variation in the end result, some tiny modification in the shape, size, or behavior of the living thing. Most of the changes tend to be detrimental and their owners disappear, but every now and then a change takes place which makes the organism better suited to some aspect of its changing environment. It survives at the expense of others less well adapted and becomes more numerous, until it in turn is replaced by a more beneficial change produced by yet another small mistake in replication.

This is how natural selection works. Change takes place and the fittest survive. And because there is an enormous range of possible adaptation, there comes to be an enormous variety of adaptive organisms. Our present richness and abundance are a result of a long and complex natural history. But in the midst of this dynamic movement, partially concealed by its very abundance, there is one strange exception. Everything changes but the replicator itself. Natural selection works on everything but the common central biochemical machinery, everything but DNA. This is why it still is common to all living things, and why it has been assumed, with authority, that DNA was there right from the start.

Cairns-Smith, however, points out that DNA is a rather fragile molecule on its own, very easily broken by mechanical stresses, even those as simple as stirring it in solution. It is also very easily damaged by ultraviolet radiation, which is certainly very powerful in space and must have been prevalent on the earth before enough ozone was formed to produce our atmospheric sun filter. Therefore DNA, if it existed at all, was a most unlikely first choice for genetic material.[92]

The first simple replicators, the most primitive genes, must have had several specific properties: They had to be self-forming under simple conditions; they had to acquire and hold information; they had to be able to use this information in some way on their environment; they had to be able to replicate the information; and they had to be able to develop and evolve.

The last two requirements are almost contradictory. Replication means making an exact copy of yourself, identical, unchanged in any detail; and evolution, by its very nature, demands change. But it is clear that no system can last long in a dynamic environment if it goes to either extreme. If it is totally inflexible, it will soon be worn away; and if it changes too much, it very quickly collapses. What it requires for dynamic stability is a capacity to change to meet new demands, and yet remain unchanged for long enough to capitalize on advances already made. It must replicate with great, but not total, accuracy. It must have a vital flaw.

DNA now has this capacity, but if it was not available in the beginning, then something else must have served in its stead. There must have been some solid-state mechanism capable of doing all the necessary things. Cairns-Smith points out that only one state of matter in the early unevolved world could have had all the necessary qualifications. He suggested that life evolved through a process of natural selection which began with the substance of the planet itself, with the dust of the earth. He is saying, in effect, that if we go in for ancestor worship, then we ought to begin to think a little more kindly of crystals.

Crystals are vivid examples of the capacity of matter to organize itself. They are regular geometric forms which seem to arise spontaneously and then to replicate themselves in a stable manner. Some substances crystallize easily, some can be persuaded to do so only with difficulty, and some, it seems, may never form crystals under any circumstances—though our inability to get them to do so may be merely the result of our limited imagination.

For example, two hundred and fifty years ago glycerine was first extracted from natural fats in the form of a colorless, sweet, oily liquid and put to use in medicine, lubrication, and the manufacture of explosives. Despite supercooling, reheating, and all the usual aids for inducing crystallization, glycerine remained resolutely liquid, and it was assumed that the substance had no solid form. Then, early in this century, something strange happened to a barrel of glycerine in transit between the factory in Vienna and the regular client in London. "Due to an unusual combination of movements which occurred, purely by chance, in the barrel," it crystallized.[410]

The client was probably livid, but chemists were delighted and began borrowing bits from the barrel to seed their own samples, which rapidly solidified in the same way at a temperature of eighteen degrees Centigrade. Among the first to do this were two scientists interested in thermodynamics who found that soon after their first crystals arrived in the mail and were used successfully for inducing crystallization in an experiment on one sample of glycerine, all the other glycerine in their laboratory began to crystallize spontaneously, despite the fact that some was sealed in airtight containers.[200] They reported this occurrence as a casual, unimportant aside in a technical paper on another topic, but today similar unintentional metamorphoses have taken place in many parts of the world and glycerine crystals are common.

This is a regular experience in organic chemistry. Yesterday something was impossible and today it is easy—partly because of the introduction of a new technique, but also in part because of the existence of a new state of mind. Glycerine crystallizes in the presence of a physical seed, but it seems that at least part of the new facility with which the process occurs is due to the presence also of an altered attitude—a kind of mind seed. Both seeds are new to the world and present something of a chicken-and-egg problem. It seems that the crystal appeared first, spontaneously, and that the idea evolved from it; but as we shall see later, it could have happened the other way round.

The evolution of life presents a similar problem, and may have followed the same kind of sequence, beginning with the existence of a suitable crystal, probably a very small one, relatively insoluble in water. A colloidal mineral would be ideal, and none is in fact more common, or better suited to the needs of a primitive gene, or more appropriate in a biblical sense, than clay.

In the beginning "there went up a mist from the earth, and

watered the whole face of the ground." Rocks at the surface began to weather and dissolve, producing dilute solutions of silica which trickled away and percolated down through porous beds, finally gathering into supersaturated solutions that, from time to time, crystallized out as clay. The synthesis of clay is very slow and still poorly understood, but the end product is very different from the feathery silicates such as asbestos and the three-dimensional-framework silicates like feldspar that make up most of the solid material of the earth's crust. Clays are extraordinary, layered, crystal structures which have, built into them, what amounts almost to an innate tendency to evolve.[558]

Most crystals occur because they become inevitable. Ice, for instance, has to happen. The characteristic arrangement of water molecules in ice is the best possible compromise between the energy in them and the tendency of all systems to become disorderly. It is the ideal state below a temperature of zero degrees Centigrade for all the molecules involved, and, until it happens, they buzz about trying out various arrangements until they find the perfect one, the one with maximum order and stability in which all the units are organized in a way that tends to lead to their continued existence. In this sense crystals display a high degree of control, but orderliness is not enough. If it were, then "the conversion of Lot's wife into a pillar of salt should be regarded as the most dramatic evolutionary advance of all time."[92] The pillar of salt was infinitely more orderly than Mrs. Lot, but not necessarily more highly evolved. Salt is extraordinarily unambitious; but clay, though equally inorganic, is a different thing altogether. Clay has plans.

Clays have a dramatic ability not only to grow, but to absorb other molecules, and this capacity varies according to their structure. Cairns-Smith constructs a possible evolutionary situation in which different patterns of crystallization in a bed of clay have given rise to three species he calls Sloppy, Sticky, and Lumpy.[92] All three types exist in a rocky stream down which trickles their "food," a flow of clay-forming solutions. Sloppy is at first glance the most successful species, maintaining a loose open structure that traps lots of food. He grows quickly and spreads out over a wide area—until it rains, and then he vanishes forever. One environmental push and Sloppy becomes extinct. Sticky is more selectively acquisitive and has picked up from the stream, along with his food, a number of sugarlike organic molecules that just happen to have been formed nearby, or to have arrived on the latest comet. Because of them he is able to stick to the rock bed and resist being washed away in the storm, but possession of these tacky molecules also restricts his size and sphere of influence, and he grows very slowly confined to that single spot—until the stream changes course, and then he dries up and blows away.

Lumpy is the lucky one. He has picked up a collection of organic molecules that make him slightly sloppy and slightly sticky, giving him the consistency of a badly made custard full of local coagulations. So, when it rains, parts of him break away and go floating off in lumps which reestablish themselves down the stream and start new outgrowths of little Lumpies in deeper pools that remain unaffected when the stream is diverted. Lumpy survives selective pressures because his adaptation makes him better suited to local environmental fluctuation than related species of clay.

"Survival of the fittest" is in fact a motto even more appropriate for crystals of clay than it is for more fully fledged organisms. Birds and beasts are not particularly adept at surviving. Individually they are poor survivors, not nearly as good as stones, for example. "A grey rock," said John Ruskin, "is a good sitter." That is its survival trick, and there are a lot of stones around to prove how effective it is. Sitting is a perfectly good piece of behavior, every bit as distinctive as the frantic dash of a dragonfly. Yet we call one alive and the other not, which is misleading. "To make life a distinction between them," observes Sir Charles Sherrington, "is at root to treat them both artificially."[486]

A dragonfly's trick is to move and multiply. And if in some places they turn out to be as numerous as stones, that is because they reproduce. They make others, again and again. Lumpy is in many ways an intermediate between stone and dragonfly. He multiplies and becomes a crowd of clay in which sheer volume and density become new environmental factors. Interaction becomes possible between adjacent lumps. One thing that can happen is that foreign atoms such as aluminum sometimes become built in among the usual silica in a molecule of clay. Each one of these then produces a local negative charge which builds up into a potential which can, if it becomes large enough, bring about major morphological changes. Under this kind of electrical persuasion, the clay might even produce a regular and predictable fold or roll in its construction, something peculiar to, and consistent with, that particular atomic aberration.

Add just one more feature to the system—the fact that a developing crystal pattern will automatically duplicate such an innovation—and you have in a bed of clay everything necessary for the acquisition and inheritance of new characteristics. You have an evolutionary system governed by a genetic mechanism.[94] At this inorganic level, Lamarck was right, acquired characters can be passed on; but this is true largely because crystals don't reproduce, they replicate. They produce exact copies of themselves, carrying on whatever information they happen to have picked up. They are stable, they replicate with accuracy; but if any change is wrought in them by an environmental pressure, they copy this imperfection as well, incorporating it faithfully into their memories.

Information in a bed of clay is carried by individual crystals and by the congregation as a whole, and it is held in much the same way as music is carried in the grooves of a phonograph record. No intelligence is implied or required. The crystals are no more "clever" than the plastic of a record is "musical." Any musical properties that a record may be made to reveal depend on the plastic's ability to accept and retain the pattern impressed on it during the recording. The memory of clay is manifest only in its ability to hold a pattern and to influence its environment when treated in a certain way.

But this ability and the pattern are vital. The American chemist Armin Weiss has shown that some clays, in particular the mica types, can build up patterns of organic molecules between their silicate layers. He has identified more than eight thousand different derivatives in which the clays have acted as templates, inducing ammonium ions and alcohols to solidify into organic components.[577] And as Cairns-Smith is quick to point out in honor of his crystal ancestors, "Reactions occurring in such an array containing suitable monomers could give rise to polymers with a genetically controlled configuration, out of which secondary control structures, membranes and other cell structures could be formed."[91] And then, as more and more of the information in the silicates was transformed to the organic molecules, the clay would cease to control and take on a more passive role as a protective clamp. Cell walls could indeed evolve at a later stage from "a vague tendency for the outer edge of a community to thicken like cold porridge, to the highly sophisticated ion and molecule filters that guard the borders of the modern cell."[92]

This scenario suggests very strongly that modern biochemical uniformity is the end product of evolution and not its jumping-off point; that protein, which we tend to regard as the be-all of life, may be only a makeshift material that was chosen in the first place just because amino acids happened to be around. Modern proteins are incredibly versatile, producing structures as diverse as porcupine quill and egg white from the same ingredients, but it looks as though they may have inherited their most important attribute from ancestral clay. Life is not in protein, but in the music written on it; in its ability to recognize other molecules and to hold ordinary atoms in an extraordinarily precise way. This, if the crystal theory is correct, is precisely what clay learned to do, and what it taught to the first complex replicating molecules cradled in the folds of its bed.

It begins to look as though the very first organism in our life system was the earth itself, in whose body developed a virus, a new metabolism drawn from ingredients in the environment, that eventually learned how to live in greater independence of its host. The whole concept of Mother Earth shifts from symbol and myth to a dawning realization that every single one of us has feet of clay, and that we live on a parent, not a planet.

There remain many formal objections to the clay concept. Supporters of the classical chemical theory are not going to abandon their position easily, but it is difficult to ignore the fact that the new idea is strongest in just those areas where the old one was vulnerable. Nobody has ever satisfactorily bridged the great gap between amino acids and fully functional, self-replicating DNA, even in the most sophisticated and carefully controlled simulations of an early world. The appeal of crystals as primitive genes is that they already possess the magic quality of replication and, given this, the gradual development of nucleic acids from simple organic substances under their control becomes a much more simple and logical construct. And simple solutions tend to be the right ones, the ones most likely to have been chosen by evolution from the comparatively simple set of conditions that existed on the primitive earth.

The famous "primitive earth" experiments, in which organic molecules are produced in spark chambers, are usually taken as evidence for chemical evolution here on earth. But, by showing that the building blocks are surprisingly easy to make, they suggest even more strongly that it could have happened almost anywhere. And our new knowledge of interstellar chemistry shows quite clearly that it did. The universe is filled with vital ingredients. There are great cosmic clouds of seeds drifting around, just waiting to take root in the right kind of soil.

It would probably be wrong at the moment to talk about organisms out there; we have no concrete evidence that evolution has gone that far anywhere else, though it would be surprising if it hadn't. But we can, with justification, begin to think in terms of complex organic molecules at large, some of which may even be at a sort of protovirus stage which could "infect" another kind of system. The early earth itself would have been a suitable host, a place with moods and rhythms of its own where, compounded with our clay and guided by its orderliness, the organic matter could be programmed in the now familiar way.

We have, in a sense, come back to a geocentric theory. Life on earth begins again to look rather special, something perhaps sparked by factors common to the cosmos as a whole, but grown in our own soil to become unique to this planet. If the cosmic seeds have taken root anywhere else, the likelihood is that their growth has been very different. The chances of the same molecules falling into the same substrate and developing the same fundamental polymers are about the same as insulin getting together by accident—negligible.

Protein is formed as a result of instructions carried by DNA, and DNA is an earth product, made of clay and manufactured here for home use only. It is the basic unit of all earth metabolism, but we must not make the mistake of assuming that is has always occupied that position, nor that it always will. It is not protected from local change, and it is not isolated here, free from cosmic influence.

There is at the moment great concern about the possible effects of research into recombinant DNA. Many fear that by tampering with genetic material, we may loose a monster on an unsuspecting world. That could happen, but we need to be equally cautious about importing life systems from other worlds. There is little chance that a true alien organism, if any such things exist, would be sufficiently like things in our system to become a local predator or parasite. But there is no guarantee that our fundamental organization is anything like the best possible solution to the problems of this environment. Other, more simple, genetic codes could have evolved elsewhere and, given a toehold here, could take over completely. If Hoyle is right about the extraterrestrial origin of epidemic disease, then we could already—or is it perhaps still?—be involved in cosmic skirmishes. Our own DNA is itself presumably the winner of just such a competition, the survivor of a battle for control, which may still be going on.

Cancer begins in a cell that divides in defiance of the body's instructions. It is in effect a parasite formed from our own tissues, a fifth columnist that is now alienated from its neighbors and no longer plays a part in the integrated community of cells that make up a healthy organism. The cancerous cell reproduces itself quickly and faithfully, just like a modified crystal, repeating the new pattern until its descendants outnumber local cells and replace them altogether. And there is good evidence to suggest that these altered instructions are imposed on a normal cell by outside agencies; by carcinogens, many of which are substances in the soil.

Lung cancer in Wales is very much more common in men who have worked in quarries producing fine-grained slate.[514] Tumors have been artificially induced in rats, mice, and guinea pigs by inhalation or by injection into the lung cavity of fine asbestos dusts.[566] The incidence of stomach cancer in Cheshire and Devonshire is directly associated with certain houses and with the soil around them.[515] One inference that can be drawn from studies such as these is that cancer, the result of an alien redirection of normal growth processes, occurs readily in organic systems that are brought back once again into close contact with a template of clay. Farmers who work with their hands in the silica of the soil and miners who inhale the silica in asbestos and shale increase the content of this simple clay in their tissues and are far more likely to suffer from cancer than their wives or others who live in the same areas breathing the same air and eating the same food.

The primitive earth is still there exercising its patterning effects, and sometimes, when it catches our DNA with its guard down, further modifications occur. Cancer is just one example of change still taking place. It is hard to ignore because of its dramatic, often lethal consequences, but it seems likely that there must be many other ways in which we continue to be influenced, powerfully and often beneficially, by our ever-present ancestor. We are embedded in the earth. "Mankind," in the words of a recent graffito, "has an incestuous relationship with Mother Earth."[16]

This is most apparent in mining, which is in essence a violation of the mother, and is in consequence accompanied everywhere by the most elaborate rites and superstitions. It always has been.[161] The oldest mine so far discovered is the one at Lion Cavern on Ngwenya Ridge in Swaziland. Material found there exceeds the limits of radioactive carbon dating and has been estimated as being perhaps more than one hundred thousand years old.[43] The shaft intrudes over forty feet into the body of earth and was sunk in search of a highly symbolic blood-red ore. The age of the mine is astonishing, more than trebling the antiquity of modern man, and showing that his genesis and formative years were spent in Africa, and not in some mythical Middle Eastern Eden. But perhaps the most fascinating feature of this ancient mine is that once the miners got what they wanted, they painstakingly filled in the excavation again, putting more than a thousand tons of earth and rock back where they found it.[71]

Earth pigments seem first to have been widely used during this Middle Stone Age period in Africa where there arose a sudden and intense demand for several minerals. Most sought after was hematite, an oxidized earth containing iron that leached upward into old beds of clay. In one form it is soft and red and usually known as bloodstone or ocher; and in another it glistens with blue-black flakes of mica, is greasy to the touch, and is known as specularite. Both earths have strong magnetic properties superimposed on the old clay templates, and it is tempting to assume that, in addition to their striking appearance, they appealed to early man because he was sensitive to them. Perhaps he recognized in these earths some trace of an ancient pattern which struck a familiar chord, and maybe he responded equally strongly to the new message inherent in them. Whatever the reasons, these soils have borne some extraordinary fruits.

Their use has never been casual. Kalahari Bushmen today still cross deserts to find specularite to rub into their hair. And in Australia, coastal aborigines are known to mount major expeditions of seventy men and more to cross hundreds of miles of hostile outback to sacred ocher sites laid down in the Dreamtime, which could be mined only by initiated old men, who crawled to the site on all fours.[4] This use of earth in adornment is one of the first pieces of evidence we have to show that early man possessed an awareness of things outside the sphere of those necessary for everyday survival. It seems to have preceded, perhaps even to have given rise to, our consciousness of mortality and the practice of burying our dead with ceremony.

One of the earliest known burials in the northern hemisphere is that of an old man, a severely arthritic Mousterian Neanderthal who died forty-six thousand years ago. He was buried in a cave at La Chapelle-aux-Saints in what is now southern France, and his body was packed around with red ocher. Now, from Border Cave in Swaziland not far from the earliest mine, comes evidence of the oldest known deliberate interment anywhere[44]—the skeleton of a child who died about eighty thousand years ago and was buried not only with a perforated seashell, probably a pendant, but also with a dusting of ash and ocher.[115]

Implicit in every funeral practice is the assumption that death is not the end, that it marks some kind of transition.[572] Early man must have noticed that the transformation occurred very often with loss of blood, and it can't have escaped his attention that women, who were monthly producers of blood, stopped doing it whenever they were involved in creating new life. Blood and life soon came to seem inseparable and led to a belief in the power of blood to regenerate and revitalize. This concept was implicit in all sacrifice and still is embodied in initiation, in the consecration of wine and in sacramental rituals such as the Eucharist. It seems likely that hematite in early ritual, and the earliest we know of was ceremonial burial, was chosen as a blood surrogate. An Australian aboriginal legend of the Unthippa women tells how they "caused blood to flow from the vulva in large quantities and so formed deposits of red ocher."[71]

Very soon the use of ocher became pandemic. The thirty-five-thousand-year-old "Red Lady of Paviland" can still be seen in the University Museum in Oxford encrusted in red ore which "colored the earth for half a yard around." Solutrean burials in Bavaria twenty thousand years ago were made by surrounding the body with mammoth tusks and submerging the entire structure in a mass of red ocher. At the earliest-known town of Catal Huyuk in Anatolia, those who died eight thousand years ago were left in the open until their bones had been picked clean by vultures, then the skulls were enshrined after painting them with hematite and cinnabar. Similarly, the Indians of North America, who were only red when painted, put their dead on platforms after making up their faces in a heavy layer of ocher. The funeral chambers of the Shang dynasty, the cists and sarcophagi of Etruscan and Roman tombs, are all painted red. Homeric dead and even the head of the Catholic Church today are buried in crimson shrouds.[71]

Raymond Dart, looking at all this evidence, was moved to remark, "Haematite has a fantastic cultural evolutionary history beginning with Mousterian burial ritual and extending through its manifold late palaeolithic artistic, religious, trading, and bartering applications. By means of its dominating agency in the diffusion of the myths, rites, and mysteries of ancient metallurgy and alchemy, it has played parts of such continuity and expanding diversity as to have rendered it unique amongst all minerals in moulding mankind's existence then and today."[69]

The enterprising and unorthodox anthropologist Adrian Boshier has discovered a telling modern example of the cultural effects of red ocher.[70] For many years he lived with the Bakgatla people of the western Transvaal and was eventually apprenticed to an old Kgatla witch doctor. His training progressed well until it reached the final stages, when time and again the necessary rituals and observances were cut short or had to be omitted because something was missing. Boshier assumed at first that the formulae had been forgotten or that parts of the ceremony had been lost through being allowed to lapse for too long. But his teacher insisted that all they needed was "the blood of the earth." Patient inquiry eventually elicited the information that this commodity was once brought to the Bakgatla by traders from the south and east, but that these men, "tall and other coloured," had long since stopped coming.

Boshier began to plot the line of the old trade route and after years of detective work, traced it three hundred miles away to Swaziland, to the ancient hematite mines which the Swazis call *ibomvu*—"the holy red." He loaded up a couple of tons of red ocher on the back of a truck and reopened the trading route by carrying the ore to the western Transvaal to give to his people. In fact he opened up a great deal more than that. Within hours of receiving "the blood of the earth," the Bakgatla were busy using it to revive customs not seen for generations. The ocher was daubed on pots that couldn't be made or used without it; the pots were involved in the ritual preparation of foods and potions for ceremonies that couldn't take place without them. Rapidly the circle spread, involving the entire tribe in the performance of complex social patterns of celebration and renewal, all made possible by the presence of a load of colored clay.

Earth, it seems, may not only have cradled life at its inception, but continues to guide and control, carrying information and acting as a cultural catalyst.

It was Boshier too who discovered why the old Swaziland mine and many others like it have been refilled.[520] The modern Swazi still, usually in secret, mine the old workings for hematite, which they smelt into iron. They believe that any digging disturbs a plumed serpent, greatest of the underworld spirits, and while one is comparatively safe on the surface, in the tunnel of a mine there is great danger from it and all spirits who resent human intrusion. So, to avoid upsetting the spirits unnecessarily, they never mine deeply and never start a new mine without first making appropriate sacrifices. During the work, daily offerings of water, meal, and tobacco must be made, and on no account must ore be smelted on or near the mine, as this would be a grievous insult. And most important of all, when work is done, the damage must be repaired by filling in all shafts with the rubble taken from them. The skirts of earth must be decently rearranged.

The French scholar Mircea Eliade points out the hidden symmetry between metallurgy and obstetrics.[159] In Egyptian, the word *bi* means both "vagina" and "gallery of a mine," and in many cultures the ore extracted from a mine is likened to an embryo and the furnace to a womb. Widespread imagery of this kind repeatedly stresses the concept of Earth as a giant mother, a being to be treated with respect. Viewed in this light, mining is a very bold undertaking and any kind of agriculture only marginally less sacrilegious.

The Sioux prophet Smohalla said, "You ask me to dig in the earth. Am I to take a knife and plunge it into the breast of my mother? You tell me to dig up and take away the stones. Must I mutilate her flesh to get at her bones?"[383] A great many myths mention stones as the bones of the Earth Mother. The implication is that, in the beginning, the earth gave birth to all beings and that the first men lived for a certain time in the body of the mother, deep in the earth itself. Still today in some parts of India tradition demands that the ill and the aged be regenerated by burying them in a grave shaped like a womb.[160] In many languages man is actually named "the Earth-born" and the belief is that babies come from earth and stones in caves or fissures. Each village, each area, knows of a particular rock or shrine that ensures fertility by "bringing" children. Even in many parts of Europe it is still possible to find a local witch or midwife who will guide hopeful mothers to the spot.

In a sense, a biological mother does no more than receive a child. And, if she lives close to her roots, she will in many countries squat or kneel and give birth directly on the ground, into the arms of the Great Mother herself. In Scandinavia, Germany, India, and Japan, there are rituals which involve placing a newborn infant on the earth. It is for Her to judge success, to say whether the birth is valid and can be taken as a regular and accomplished fact. And when such a natural man dies, the great desire is to return to the Mother, to be interred in the "native soil." This accounts for the widespread fear of being buried anywhere else, and for the obvious satisfaction expressed in many epitaphs—"Here he was born, and here he died."[159] Dust to dust again.

If the earth is the mother, then the father is usually seen as a spirit, something insubstantial passing by. He may be a plumed serpent like Quetzalcoatl, godhead and symbol of Mexico. Or a multiheaded snake like endless Ananta, the Hindu keeper of life energy, guardian of the portals of earth.[599] Or perhaps as a legless dragon or something like the Great Worm of Spindlestone.[134] But in every case, though he may be manifest in the underworld of earth, though he may even be "The Snake That Supports the World," his existence here is fleeting and he owes his true being to the sun. To Mitra, inseminator of our planet. Mother Earth and Father Sun.

The Zuni Indians call themselves "Children of the Sun," and in their creation myth, the "first beginning according to words," they tell that "in the beginning of the new-made, the All-container conceived within himself and thought outward in space, whereby mists of increase, steams potent of growth, were evolved and uplifted . . . he made himself in person and form of the Sun whom we hold to be our father . . . and with the brightening of the spaces the great mist-clouds were thickened together and fell . . . impregnating the world-holding sea."[549]

It would be difficult to produce a scientific description of the universal big bang, the accumulation of interstellar clouds under gravitation, the coalescence of cosmic dust into stars, the formation of planets and the birth and arrival of a comet, in more precise terms than those. And impossible to design a way of putting life and man more elegantly, more lovingly, into cosmic perspective.

Those "mists of increase" have been called many things. They are "Kundalini," "serpent current," "telluric force," and "subtle energy," but all in the end speak of the same thing. And all seem to apply without effort to our new myth, born of Hoyle and Cairns-Smith, telling of serpentine strands of organic molecules drifting down out of interstellar space to infect, vitalize, and be organized by the crystal clays of earth.

And the earth spirit, its name is legion too. It is the prana or mana of Eastern metaphysics, the vril or universal plastic medium of occultists, the anima mundi of alchemy. Its modern discoverers have each given it names of their own like animal magnetism, odyle, and orgone energy. Wilhelm Reich tried to catch it in boxes of wood and metal, and August Strindberg hunted it with a bottle of liquid lead acetate in the cemetery of Montparnasse.[375] But it is at heart a simple thing, something down-to-earth, perhaps no more than a rhythm inherent in the laws of thermodynamics that allows our fabric, the soil itself, to exist in an orderly way. "As if this earth in fast thick pants were breathing" life into the dust of the ground.[114]

# CHAPTER THREE

# THE FLOWER: Evolution

Each of us is a mobile museum. The fluid in our bodies is a perfect replica of that ancient sea in which we grew to fruition following our liberation from the clay. The concentration of sodium, potassium, and chloride in our blood, the cobalt, magnesium, and zinc in our tissues, are the same as those that once prevailed in the primordial ocean.

We still carry that ocean around inside us, trapped there like a living fossil forever. And in each miniature internal sea, the same old struggles go on much as they did three billion years ago. Competition and cooperation, occasionally even outright war, take place among arrays of organic compounds, and only those with the greatest physical strength or ingenuity, the most resolute chemical tenacity, survive.

That's how it is, and that is how it has been for a very long time. The compounds which make up our metabolic pathways, the sugars and fats that transport energy, the amino acids that construct our proteins, are the best that time can buy. They are the winners in a struggle for survival almost as old as earth itself. And at each stage of the struggle the victors, those with some new trick or twist, have been able to preserve their advantage for future use because of a vital development—the first protection racket.[206]

Without protection, newly formed compounds would have gone on simply replacing one another until all, winners and losers alike, were washed away by the tide. There was no real evolution, interstellar or terrestrial, until some system of protection had arisen to keep hard-won gains from loss and disintegration. At first it was probably very simple—perhaps just a collection of dust grains that offered some shelter from cosmic radiation. Later maybe there was a shield or barrier to cover those parts of a new polymer most sensitive to heat or damage. This function could well have been served by special configurations in a bed of clay; but finally, when the molecules learned to replicate in their own right and went out into the world, they had to develop something more adaptive, more elastic: a membrane.

Without a membrane, a cell wall, little further development

would have been possible. Without protection for all the marvelous machinery, there would have been no more complex assemblies, no refinements, no growth through trial and error, no momentum on which to build. The wall isolated a small part of that early ocean which formed as antique mists continued to "water the face of the ground." It separated a tiny spoonful from the rest, an area of much more manageable proportions, where relative security prevailed.

Within these confines, the early polymers and enzymes were assured of some continuity. The ancient seas, exactly as they were at that moment, were suddenly and permanently preserved. By the simple expedient of cutting a minute part of that ocean off from the main body, it and its contents were protected forever from all the environmental changes which have taken place ever since. Ice ages have come and gone, lakes have silted up and formed again, even the air has changed; but those ancient seas persist. And in them, there is an archetypal being, the first replicator; still running things, still making careful arrangements to ensure that the walls are maintained for generation after generation. Alive and well and living in every cell, our founding fathers, those immortal coils of DNA.

At first the earliest replicators were exposed, like everything else, to the hurly-burly of open competition. But then came the walls, and they were protected in their private pools by barriers of protein. Every pool had a replicator. In fact, the conclusion seems inescapable that it was they who invented the cell walls, who designed and built the barriers for their own protection. They constructed vehicles for their continued existence, survival machines for themselves to live in.

The first such vehicles probably consisted of nothing more than that membrane, but in time and in response to various environmental pressures, the machines became more and more elaborate. Until today, four billion years on, we have the whole panoply of evolution. And what of those early architects, the replicators themselves? As ethologist Richard Dawkins sees it, "they did not die out, for they are past masters of the survival arts. But do not look for them floating loose in the sea; they gave up that cavalier freedom long ago. Now they swarm in huge colonies, safe inside gigantic lumbering robots, sealed off from the outside world, communicating with it by tortuous indirect routes, manipulating it by remote control. They are in you and me; they created us, body and mind; and their preservation is the ultimate rationale for our existence. They have come a long way, those replicators. Now they go by the name of genes, and we are their survival machines."[125]

It is very difficult to get used to the idea that we may be shared, rented, and occupied in this way. It goes against the human grain. The paragon of animals nothing more than a taxi for a bunch of chemicals? Ridiculous! And yet the evidence is strong, and direct. We die, but the gene doesn't. It never even grows old, but leaps from body to body down through all the generations, manipulating each in turn for its own ends, abandoning a long succession of convenient vehicles along the evolutionary highway, where they lie rotting until another genetic passenger comes along and scavenges the spare parts to build itself a better gene machine.

This genetic viewpoint was crystallized a long time ago in Samuel Butler's famous observation that "the hen is only an egg's way of making another egg."[113] The body is only a gene's way of making another gene. An organism does not live for itself. Its primary function is not even to reproduce other organisms like it, but to replicate genes for which it serves as a temporary carrier.

What about free will? Where does the individual figure in all this? Am I not unique and possessed of a particular destiny? Yes, up to a point, but the life of even the greatest of individuals is short, and posterity is totally confused with that of one's mate. My children are only half me, and my grandchildren no more than twenty-five percent. The best I can hope for is a large number of descendants, each of whom bears a tiny portion of me and equally tiny parts of a great many other people. There is no future for an individual. We are just fleeting things. Even our chromosomes, the hand we have been dealt and of which we are so inordinately proud, are evanescent. They get shuffled into oblivion with each new deal. Only the cards themselves remain unchanged. The cards are the genes, and genes are forever.

There is even a sense in which the gene denies an individual any real existence at all. Your body contains about sixty trillion cells, and every single one of them holds a selection of genetic material, a complete blueprint for making you again, correct down to the last detail. This set of plans is called a genotype, and the expression of the instructions contained there, the individual organism, is known as the phenotype. It is difficult to deal with anything as complex as a human in simple terms, so imagine that the phenotype which concerns us is a stone, an ordinary sort of oval pebble lying in the bed of a mountain stream. The environment in which the stone exists is represented by the water flowing endlessly by. Imagine too that the characteristics of the stone are determined by a pattern of atomic energy that holds its matter together in a particular shape. That pattern is the genotype, the plans for that stone. We cannot normally see the genotype. All that is apparent to us is the shape of the stone, its outline, which is marked by points where smooth-flowing water is thrown into confused eddies by the effects of something in its path. But that is only part of the story of the stone. There are all kinds of things like shock waves, temperature and pressure gradients, and energetic exchanges going on through the fabric of the pebble that are invisible and perhaps even impossible to measure in any way. All we are left with is a superficial perception. The shape of the stone, its phenotype, can be considered in fact as nothing more than a convenient term for describing a rather ill-defined part of the environment, that part where it just happens that the most intense reactions between the environment and the genotype are taking place. It is a marginal area, a place of flux and change where nothing is certain and reality exists only from moment to moment. T. S. Eliot said, "What is actual is actual only for one time, and only for one place."[163] The individual, if it can be said to have any reality at all, is of but momentary significance.

This is perhaps an extreme view, which applies equally to each individual molecule of DNA, whose life can be measured in months. But DNA is the replicator, making exact copies of itself as often as necessary, and can theoretically continue to exist for five billion years or more. It is limited only by its inventiveness in designing vehicles in which it can survive under any circumstances. As Dawkins says, "A monkey is a machine which preserves genes up trees, a fish is a machine which preserves genes in the water; there is even a small worm which preserves genes in German beer mats. DNA works in mysterious ways."[125] But we are beginning to understand the rules of the game.

For a start, it makes no sense to think of the genes as conscious entities. They have no direct knowledge of us or of each other. They do not know they are involved in evolution, they do not make plans for future development or dream up new survival strategies. They just exist. But they are highly gregarious, getting together in thousands to direct one of their machines to a particular goal. And they succeed because they operate according to the principle of natural selection.

A great deal has been written about natural selection since Darwin and Wallace drew attention to it in 1858, but it is at best only imperfectly understood. It is based on the fact that variation exists within any given population of organisms; and on the assumption that some of these variants will survive at the expense of others. They will be better adapted to conditions and therefore more likely to breed successfully. For this reason the successful variants are said to be selected by a natural process that ultimately produces an evolutionary change, a lasting modification in the species. But this bald statement, which is as far as even most biologists feel the need to go, tells nothing of the sophistication that exists within the mechanism, or of the delicate checks and balances which it exercises. One has to see it in action.

High among the Retiche Alps of Switzerland, in the upper reaches of the valley of Engadin, there is a superb forest of larch. Normally these conifers rise in pale-green pyramids right up to the timber line, but once every seven or eight years their canopies are burned brown and bare by a population explosion of *Zeiraphera griseana*, the gray larch budmoth. For the past twenty years, local biologists have been involved in an intensive study of the moth's life history and have recently made a fascinating discovery.[111] The eggs hatch in May and the larvae feed on the needles of the larch, going through five growth stages in about fifty days. Then they drop to the ground, where they pupate on the forest floor, emerging as adults thirty days later. The gray-brown moths live for five weeks, finally dying in late September after laying eggs on the bark of their host trees under protective patches of lichen.

Nothing peculiar in that cycle, but what makes it extraordinary is that the species exists in two distinct forms which go through the whole series of events in the same piece of forest at the same time. The two forms are called "Weak" and "Strong" and can be distinguished by their metabolic rates. The Strong variety produce caterpillars that thrash about vigorously, eat more, travel farther, and grow larger than their Weak cousins. The Strong moths fly more readily, and lay more eggs with a higher rate of success in overwintering and hatching.

By all the rules of natural selection, the Weak variety should long since have ceased to exist, but the researchers have found that they harbor a secret weapon. Their less exuberant metabolism is accompanied by a built-in resistance to a granulosis virus which is carried in latent form in the bodies of all members of the species. And once every seven years, when the Strong types have succeeded so well that the larches are turning brown, population densities produce unusual pressures on the insects and the vigorous moths succumb to stress and virus disease, and nearly all of them die.

Then the Weak inherit the earth of Engadin and come into their own, but they too are kept in differential check by a parasitic wasp which lays her eggs on the backs of moth larvae which in the Weak variety are not strong enough to defend themselves. So the cycle starts all over with the higher reproductive potential and greater dispersal abilities of the Strong types gradually reasserting their superiority in terms of numbers, until the explosion and crash occur again.[30]

The point of the story is that this remarkable cooperative situation could never have come into being if natural selection were concerned only with the survival of the fittest individuals. One or both of the forms of this budmoth would simply have ceased to be. But evolution does not deal with survival machines; it is concerned with, and ultimately governed only by, the selfish interests of the genes responsible for the species as a whole. And as long as the genotype is preserved, those self-centered chemicals couldn't care less which vehicle happens to carry them, just as long as they continue to travel. The ultimate purpose of the genes is simply to survive.

Dawkins describes the type of control practiced by the gene as being similar to that of a program used to help a computer play chess.[125] "The chess-playing programme is not fussy which physical computer it uses to act out its skills." And once the program has been inserted, the computer is on its own. There is no further intervention except for the opponent, the environment, typing in his moves. Because there are more possible games of chess than there are atoms in the universe, it is impossible for even the best program to anticipate all likely moves. And it is equally absurd to expect the computer to try out all configurations opened up by every new position. The world would come to an end before it got even halfway down the list. So the best a program can do is to tell the computer the basic moves of the game, and to add a few hints about strategy and technique.

All the genes can do is set up a survival program in advance and then sit passively inside, leaving the organic vehicle to do the rest on its own. The genes build an executive computer for themselves and program it in advance "with rules and advice to cope with as many eventualities as they can anticipate. But life, like the game of chess, offers too many different possible eventualities for all of them to be anticipated." Like the chess program, "the genes have to instruct their survival machines not to specifics, but in the general strategies and tricks of the living trade."[125]

This is a slow and rather remote form of control, but nevertheless very powerful. The genes take a gamble with each program they write; they cannot possibly predict all the difficulties likely to face each individual computer. But the gambles tend to pay off because they are interested only in the final result. They expect to win on average at the end of a long run. So they weigh up the odds and place their bets accordingly. And the gamblers who win go on to play again in the next game. Winner takes all in natural selection.

Naturally, strategies differ very widely, depending on the stake, the odds, and the prize. If the prize is large, it might be worth risking a big investment. "A gambler who risks his all in a single throw stands to gain a great deal. He also stands to lose a great deal, but on average high-stake gamblers are no better and no worse off than other players who play for low winnings with low stakes."[125] A human male and an oyster are low-stake players in the reproductive phase of their games. Each puts out a spray of millions of tiny seeds and lets them drift at the mercy of the tide. The human female is a high-stake player, risking all her reproductive potential in the single throw of an egg. On average, though, things tend to even out, with the survival of approximately one new human and one new oyster.

Genes dictate the way in which the machines are built and the manner in which they operate. They have ultimate power over behavior. But because they might have to wait for a generation to assess the effects of their control, moment-to-moment decisions about what to do next must be taken by the individual. Genes are the policy makers, we are their executives. But as evolution progresses, the executive apparatus has become increasingly sophisticated and management has begun to make more and more decisions on its own. Nervous systems have evolved to levels where learning, memory, and model-making become possible and take over many of the policy decisions. And the logical conclusion to this trend would be for the genes to send out a very elaborate survival machine with only one all-encompassing instruction—"Do whatever you think best to keep us alive." But no species on earth has yet reached that level.

We are all, every creeping thing after its kind, to a greater or lesser extent governed by proxy—by a three-lettered deity we call DNA.

The seat of our government, the shrine of our god, is simple enough to see. At an early stage in the development of every cell in each of our bodies the genetic material and certain specific proteins coagulate into threads that come to life in a tangle near the center of the cell. The substance in the threads readily picks up chemical dyes and has for that reason been called "the colored stuff"—chromatin—and the threads are known as chromosomes. Each one holds volumes of information, precise instructions on how to build and program an organism like a man. In the same way that the cell is cut off from its environment, so this little library is isolated from the rest of the cell by a membrane. Outside is the cytoplasm, the basic matter of the cell with various small inclusions; and inside is the nucleus, the inner cabinet where top policy decisions are made.

It will perhaps help a little to understand how something as comparatively simple as nucleic acid can exert so far-reaching an effect, if we look briefly at how it came to be sitting there, at the center of things.

Put a thin overcoat, a jacket of protein, onto a strand of DNA and you have a virus, the simplest almost-living thing we can imagine. Turn a number of these replicators loose in the primordial ocean and soon you have a very great number of them. In time that is all you have, because the chances are that in their singleminded way they will use up all the available raw materials to make more of themselves. Eventually, in a world in which nothing else exists, they have to begin to take notice of each other. Three types of interaction are possible. Either they compete with each other, or they eat each other, or they cooperate. A virus can inhibit others, so that it gets to the food first. It can treat other viruses as food and develop some way of breaking them down and ingesting them. Or it can cooperate with them to gain access to new materials by producing results that neither could achieve alone. All these reactions certainly occurred in the early ocean, but the first two did nothing more than postpone the crisis, so ultimately evolution depended on cooperation of some kind—on a primitive type of living together. There are still viruses around which behave in this way.

In 1910, Peyton Rous isolated a virus from a tumor growing in a Plymouth Rock hen and found that it caused a similar cancer when injected into other chickens. This was the first demonstration that a virus could be involved in malignant growths, and it caused great excitement in the community of cancer researchers, but interest waned when it was found that the Rous sarcoma virus (RSV) failed to induce cancer in any animal but the chicken. The reason only became clear in 1964 when Harry Rubin at Berkeley discovered that in fact not even RSV could cause tumors, even in chickens.[467] At least not without help. By diluting a culture until there was no RSV left, Rubin found that there was still something there. Another virus, a hidden passenger now called the Rous Associated Virus (RAV), was creeping into the chicken cells along with RSV. On its own, RAV couldn't get in, it was equally helpless, but magnificently chivalrous. RSV could get into a host cell without aid, but couldn't reproduce there. It was chemically inhibited from producing extra protein coverings; so RAV, Raleigh-like, lent RSV its coat. Together, they were a great success in the cancer business, and now research has begun looking askance at all kinds of associations between otherwise totally innocuous substances.

Living together for mutual benefit, a process known as symbiosis, is the best possible way of producing dramatic evolutionary leaps. It involves no basic change in the genotype of either contributing partner and yet it results in an association which takes both members into a whole range of new situations with new possibilities for survival and growth. It is no accident that of all complex organisms on earth today, the group which contains the hardiest pioneers, those first to settle in all the most formidable terrestrial environments, are symbionts. They are found on rocks among the permanent snow on mountain tops, on outcroppings not far from the poles, and on sun-baked desert floors too hot to put your hand on. They have even been placed in liquid oxygen at $-183$ degrees Centigrade, or kept in a vacuum for six years, and still survived. These are the natural synthetics we call lichens.

There are more than fifteen thousand known species of lichen, each with its characteristic phenotype of a definite shape, size and color; but in a sense they are all illusions.[331] They have no independent biological reality, being composed of two distinct plants. One is an alga, a tiny sort of seaweed that manufactures its own food by photosynthesis; and the other is a fungus, a mushroom that has to live off other organisms. The two together form a third entity quite different in appearance and life history from either of its components and adapted to environments in which neither could survive alone. In fact, the fungal partner depends totally on food produced by the alga and no longer exists anywhere in nature at all except as a member of its powerful union.[1] The algae can live alone and some of them still do, but they benefit from the relationship by receiving protection against radiation, desiccation, and mechanical injury. Together, they are wonderfully resilient; some colonies reach ages of more than two thousand years, clinging to bare rock like time stains.

Botanists have succeeded in separating the two components in the laboratory and culturing them independently, but it is extremely difficult to put a lichen back together again and make it work. In 1966, Vernon Ahmadjian of Worcester University did succeed in getting one fruticose lichen to recombine by starving the two ingredients until they were forced to associate in order to survive.[2] There are in nature certain types of fungi that live in close contact with colonies of algae in a sort of trial marriage, and it may have been on associations such as these that natural selection, with the help of a certain amount of hardship, drew to produce the symbiosis.

And it seems likely that it was early pressure of just this kind on the most simple organisms that induced them to cooperate in new and more productive ways.

As with theories on the origin of the first replicators, there are many ideas about the development of the present complex plant and animal worlds. The classical theory in this case suggests that an early organism learned how to make its own food, and that this plantlike bacterium evolved gradually into the plants; while somewhere along the line some cells lost their food-making ability again and evolved into fungi and animals.

Lynn Margulis of Yale University calls this "the botanical myth."[362] She points out that mutation and natural selection can account for the step-by-step development of most modern organisms, but that there is one basic discontinuity very low down on the evolutionary tree that can't possibly be explained in this way. "Every form of life on earth—oak tree and elephant, bird and bacterium—shares a common ancestry with every other form; this fact has been conclusively demonstrated by more than a century of evolutionary research. At the same time every living thing belongs primarily to one or another of two groups that are mutually exclusive."[363] The gap in question is great and exists not between plants and animals, which in fact have a great deal in common, but between all the advanced cellular organisms and the simple group made up of bacteria and blue-green algae. Between the haves and the have-nots. Between those whose cells have nuclei and those whose cells do not.

The cells of all advanced organisms like palm trees and people are larger, have sexual systems, need oxygen to live, commonly develop into multicellular associations, and have a nucleus of DNA which is surrounded by its own protective membrane. They are called eukaryotic cells, which is a name derived from two Greek roots that simply mean "having a true nucleus." The cells of bacteria, on the other hand, are smaller, have no sexual systems, are often killed by oxygen, seldom develop multicellular structures, and have no nucleus. They are known as prokaryotic cells—those which are "pre-nuclear."[541]

Margulis suggests that in the Precambrian period, something more than six hundred million years ago, the seas and lakes, the soil and air were colonized by a variety of organisms belonging only to the prokaryotes; none had nuclei and all depended rather heavily on food produced by the common blue-green algae, which were the only ones practicing photosynthesis. As food grew short, many different kinds of relationships grew up among the microbes, and it was a succession of combined efforts that eventually gave rise to the eukaryotes. She believes that the cells now common to every advanced living thing, whether it be a sperm or a sperm whale, consist of a complex community of three of those early prokaryotes now living and working permanently together.

The three she has in mind are the nucleus, which holds our DNA; the mitochondria, which are inclusions in our cells that arrange the way we use oxygen; and the cilia, which are tiny whiplike structures that make some of our cells mobile and others sensitive in a variety of ways. In effect the three in-gredients are the modern cell's brain, its heart, and a sort of combination of legs, arms, and sense organs.

The get-together began, it seems, with simple oxygen-breathing bacteria taking up residence in stray pieces of pro-tein. On their own, neither associate was of any great conse-quence, but together they formed a directed and distinctive cell something like a modern ameba. At first their relationship was just a working one; they stayed together in order to make a living. But then the partners, the protein and the bacterium, began to redecorate their premises. They got together like the plant associates in a lichen and produced something entirely new. The idea was probably the senior partner's, because what they did was to build a wall to give his DNA a private office. This is the inner sanctum we now know as the nucleus, and the junior partner, with the title of mitochondrium, was rele-gated to a desk in the outer office where he still continues to keep an eye on fuel and energy consumption.[103]

It seems likely that this nucleated early ameba became one of the biggest businesses around, flowing through the oceans and taking over every small concern it encountered. Until eventually it incorporated another business that was too good to be used simply as a source of raw materials, and decided instead to build this intact into the system in a useful way.

Margulis suggests that this was another sort of bacterium, a wriggly little microbe like a corkscrew without a handle, and that it joined the firm in the transport department.[364] It attached itself to the outside wall like a propeller so that, instead of just oozing around, the entire concern could swim rapidly through the water in search of new and bigger markets.

If you think this whole idea is rather far-fetched, you're in for a surprise, because something very much like that early cooperative still exists.

Most plants are able to stand upright because of structural girders built of cellulose and lignin. Anything that feeds on plants has to cope somehow with these tough substances, and many herbivores simply resort to eating a great deal in order to get the small amount of nutrient they need. The caterpillar of the goat moth, *Cossus cossus*, burrows into the bark of elm trees and increases its weight by seventy-two thousand times in a period of three years, but it has to eat an enormous amount of wood to get this done. If it is taken off elm and given something more nutritious like a beetroot, it matures in a matter of months.[278] This is clearly wasteful for both moth and elm, so many species have tried to find a way round the problem. Nobody has yet discovered any way of dealing with lignin, but there are a few bacteria that have learned to handle cellulose, and many insects employ these diligent workers in fermentation factories. Termites are the best-known employers; many have turned their enlarged hindguts over totally to a whole range of symbionts who do nothing but process cellulose, which the insects provide by pulling down trees and houses. With the aid of their employees, some termites can even live on a pure cellulose diet such as blotting paper.

In the Australian termite *Mastotermes darwiniensis* the symbionts have formed a workers' cooperative that goes under the heading of *Mixotricha paradoxa*.[75] The concern is classified as a flagellate protozoan, a smooth-swimming, single-celled organism which seems to be covered with tiny hairs that beat in rhythm to move it along. But the paradoxical species name is all that protects this combine from prosecution under the Trades Description Act, because it is no ordinary unicellular animal.

A close look at the undulating oars on *Mixotricha* shows that they are in fact elongated bacteria harnessed to the surface of the colony and helping it to glide along at constant speed in a straight line through the fluid in the termite's gut, until it comes up against a splinter.[226] Then another part of the combine takes over. At the base of each oar is a hollow of exactly the right shape and size to hold another bacterium, this time an oval one, which it seems may help coordinate the movement which brings each cellulose meal around to a sort of mouth, where it is swallowed. And inside are still other bacteria, several types of them, that undertake the actual fermentation.

There may once have been a free-living piece of ancestral protoplasmic protein that warranted a name of its own, but the beast we now know as *Mixotricha paradoxa* is very clearly a commune of once independent organisms, now collected together under one roof in activity designed to further their mutual interests. And the totally astonishing thing about this, which still amazes and disturbs me every time I think about it, is that every human cell is built up in the same way, out of separate individuals which still to a certain extent continue to lead their own lives, speaking their own languages. Our mitochondria, the little engines in our cells that decide how and when we use oxygen, are built of proteins quite unlike anything else in our bodies. They are more closely related to each other, and to free-living bacteria out there in the rivers, than they are to us. We are colonized by strangers.

We are, however, not alone in this predicament.

All green plants have their own independent lodgers. They couldn't be plants or green without them. They are green purely and simply because they contain the pigment chlorophyll in little disc-shaped chloroplasts that rest in most of their cells. Looked at closely, these green particles bear such a remarkable resemblance to free-living blue-green algae that it seems fair to assume the plants may have acquired them when some ancestor at large in the primordial ocean swallowed one and has kept it ever since, hard at work in a succession of descendants, making food by eating the sun.

Unicellular algae are enormously cosmopolitan, living in seas, lakes, deserts, on rock, mud, or ice. Sometimes they occur singly, but most often in colonial masses that color whole areas in a characteristic bloom. Occasionally they take refuge inside, or are invited in to join, other organisms. The slug *Elysia viridis* eats the green seaweed *Codium fragile* and keeps some of the plant's chloroplasts going in its own tissue, so that it looks like a crawling leaf.[545] And the commandeered plant factories carry on turning out photosynthetic sugars for the benefit of their new host, who needs to forage for itself far less than other slugs that are still self-employed. There are thriving colonies of algae in some small marine worms that live along the tideline. There are others in the throat of certain kinds of clam. There are some that add camouflage to their hair on two-toed sloths. And there are many that cooperate with protozoans.

Most interesting of the latter is one little jellylike blob that glides through the water of stagnant pools fluttering its eye-lashes.[29] *Paramecium bursaria* is easily identified because it is bright-green, and like plants, gets its color from chlorophyll. This protozoan feeds on algae, having a special preference for the common *Chlorella*, most of which it digests. But, in a way we still don't understand, several hundred algae survive intact and take up residence in their host's cytoplasm. There they carry on their normal photosynthetic activities and can, when times are hard, even keep their hosts alive by making enough food for everyone. In return, they benefit when there is no sunlight, by living off the body of the protozoan.[306]

It is possible to tease the host and its guests apart and show that they live perfectly good independent lives. But if given the choice by being brought together in the same culture, they soon team up again. The host protozoan takes in guests or they breed inside, but only until the algae have reached their appropriate optimal number, then any new *Chlorella* that comes along is dealt with in the normal way as a foodstuff and digested. It is clear that the association is not just a random capricious one, but a long-term symbiotic relationship held in very delicate ecological equilibrium.

All that is necessary to bridge the gap between this sort of association in which an organism entertains a food maker and the green plant situation in which food-making particles are permanent parts of the metabolism is to show that a symbiotic partner is passed on from generation to generation in some form of inheritance.[532]

This happens often. A large number of plants, which on their own are unable to take up free nitrogen from the air, use bacteria to make this trick possible. The bacteria by themselves can't do it either, but peas and alfalfa working together with *Rhizobium* bacteria have invented a new phenotype. They build special root nodules in which nitrogen fixation can take place. These lumps on the roots actually become pink with a substance very much like the hemoglobin in our blood, and it is this that makes the new chemistry possible. But each new plant has to start all over again establishing its own association with a new generation of bacteria.[307] Some tropical members of the madder family, however, have found a way to incorporate the vital *Rhizobium* bacteria into the structure of their seeds, so that the seedlings of the shrub *Psychotria bac-*

*teriophylla*, for instance, inherit not only parental DNA, but mother's little helpers as well.

There are in present natural history all the mechanisms necessary to account for the congregation of free-living organisms, for their effective cooperation, and even for their reproduction as corporate wholes. Lynn Margulis concludes that "the mechanisms by which very different partners encounter each other, develop stable relationships with each other over time, transfer metabolic products and even genetic material from one to another, insure the stability and specificity of their association, and finally fuse so that the probability of separation and even recognition of the partners becomes difficult, are still largely unknown."[365] But slowly, step by step, even the most cautious biologists are reaching an inevitable consensus. A theory that only a few years ago was considered "too fantastic to mention in polite biological society"[591] is now gaining general acceptance.

It seems impossible to deny any longer that all eukaryotic cells—that is, the cells of every complex living being—are social structures that came about more than a billion years ago when certain big cells engulfed a whole lot of small ones that have stayed on there inside them, relatively little changed, ever since.

There is something vaguely familiar about the picture of a small cell entering a large one and staying on there to produce something new. It reminds me of sex—specifically of the way in which a tiny sperm enters a large egg and sparks off the whole process of embryonic development. Permanent stable symbiotic relationships are in a real sense "parasexual phenomena"—bringing together inside the same individual two independent sets of genes with their separate books of instruction. In true sexual reproduction, pages are torn out of each book and a new text with a new binding is released. This never happens in quite that way during even the most advanced symbiosis, but there is good evidence that at least some genetic mixing takes place during associations of even the most radically different species.[365]

When Aristotle first saw an ostrich, he is reported to have said that "this great bird can only be a cross between a mosquito and a giraffe." He probably wasn't serious, but we know that interspecific hybrids are possible, at least between closely related species. Lion and tiger (*Felis leo* × *Felis tigris*), horse and donkey (*Equus caballo* × *Equus asinus*), aspen and cottonwood (*Populus tremuloides* × *Populus trichocarpa*), all produce offspring which share the characters of both parents but are usually sterile and unable to pass them on. This is bad news for the genes, so in practice there are commonly barriers of various kinds erected between similar species to prevent them from throwing themselves away on close relatives. But there is no evidence of such taboos or restrictions preventing the association of radically different species. As far as symbiosis is concerned, the more they differ, the better. We are only just beginning to realize what evolutionary potential exists in such liaisons. The natural development and growth of orchids depends almost totally on the penetration of the root of the seedling by a forest fungus. The success of many conifers likewise revolves around a mycorrhizal fungus that forms a beneficial relationship with the roots of the young trees. It has, however, been suggested that the conquest of dry land by green plants would have been impossible without symbiotic fungi such as these to protect and nourish the first pioneers.[365]

It has never been easy to define a species, but in symbiosis it becomes almost impossible. A Belgian team of plant geneticists has shown that a variety of the mustard plant *Arabidopsis thaliana* which has a lethal hereditary defect can be cured by association with bacteria.[339] The breed is unable to produce a type of vitamin B called thiamine, and without this vital catalyst, although its seeds germinate, they very soon die. But many bacteria make thiamine, and when DNA from these is applied to the mustard seeds, they absorb the necessary genes, build them into their own genotype, and grow into flowers that set normal seeds. The genes of plant and bacteria combine permanently to form a new individual of what is in effect a new species.

This is a shattering discovery. At first few biologists believed it could be true, but the same team has repeated it with barley, and now others have gone on to show that mutations take place in fruit flies infected with certain kinds of bacteria.[182] These effects have admittedly been produced under artificial conditions in the laboratory, but evidence is beginning to pour in to show that genes in nature are not nearly so jealous of their substance and purity as we once imagined.

Many wild bacteria spontaneously release their DNA into a growth medium or into the plant cells they happen to occupy at the time.[519] And many other organisms pick up and make good use of stray DNA encountered in this way. It even begins to look as though we should consider viruses as mobile genes, intent not just on dealing out random death and disease, but on sharing news, some of which may well be good. Research at the National Cancer Institute in Maryland has shown that the chromosomes of some cats, both wild and domestic, contain genes quite unlike those in all other cats, but identical to genes found in many higher primates such as baboons and monkeys, and known also to exist in a particular virus. And they conclude that "viruses have under natural conditions transferred genetic information between species that are only remotely related."[47]

This exchange of genetic information is already throwing biologists into states of great confusion. We have always believed that a close resemblance between any two species means that they must have recently shared a common ancestor, and that the main problem was keeping these close relatives apart, so that each could continue to breed true in its own way. There is no place in classical evolutionary theory for the idea that completely different species could be reproductively involved. And until very recently nobody suspected that the similarities between widely different groups were anything more than an accident we describe as convergent evolution. But now it becomes a distinct possibility that something like the independent evolution of quills and spines in the European hedgehogs, the African porcupines, the Australian echidnas, and the North and South American porcupines—none of which has any kind of direct relationship with any of the others—may not just be fortuitous, but a consequence of "something in the air."

Everything is being turned upside down with evidence that the gene pool is very much larger than anyone imagined. Nothing is sacred any more.

One brave cell biologist has even said that "there is substantial evidence that organisms are not limited for their evolution to genes that belong to their own species . . . the whole of the gene pool of the biosphere is available."[287] We are not confined to the limits of our own inheritance. We are not subject only to the dictates of our own governing body. We, and every other living thing, are involved in an exchange of mutual interests. Science at last is beginning to take seriously the age-old mystic contention that all are as one. We are not alone.

Lewis Thomas as usual puts it effortlessly and beautifully: "We live in a dancing matrix of viruses; they dart, rather like bees from organism to organism, from plant to insect to mammal to me and back again, and into the sea, tugging along pieces of this genome, strings of genes from that, transplanting grafts of DNA, passing around heredity as though at a party."[536] Viruses may be our earth organism's transport system, a mechanism for keeping the best bits of information, the latest gossip, in the widest circulation among us. And we ought perhaps to look on the odd ill effect, the occasional virus disease, as only an accident.

There is a simplicity, an internal logic and mythologic consistency about this whole idea, that makes it irresistible. One keeps coming across new evidence that, far from being islands unto ourselves, we do our level best to keep contacts going. I have just discovered a wasp that passes her endosymbionts, the useful army of bacteria and yeasts in her gut, on to her offspring by anointing each and every egg she lays with a little dab of loaded fecal material.[319] Her genotype now presumably includes specific instructions for a new piece of behavior dedicated totally to establishing an heirloom, an inheritance which the wasp genes themselves cannot provide. This is a good example of outside aid, but it seems that we have been grievously underestimating our own internal potential.

We know a great deal more about the genetics of the fruit fly than we do of our own heredity. It might be different if humans could be raised on banana mush in little bottles and persuaded to breed every three days. But slowly we are beginning to pinpoint various functions on maps of our forty-six chromosomes and, as our knowledge grows, even the most conservative estimates suggest that there must be a million or more genes in our nuclei that we are just not using.[119] We have enormous genetic deposit accounts on which we could presumably draw in times of need. Or perhaps they are historical records, files on everything we once used to be. It seems to be true that only one percent of this archive is uniquely ours. We share the other ninety-nine percent with our nearest living relative, the chimpanzee. And he, like us, is subdivided and rented out to total strangers. We carry within us, even without the aid of peripatetic viruses, the seeds of the world.

"At the interior of our cells, driving them, providing the oxidative energy that sends us out for the improvement of each shining day, are the mitochondria, and in a strict sense they are not ours. They turn out to be little separate creatures . . . probably primitive bacteria that swam into our ancestral cells and stayed there. Ever since, they have maintained themselves and their ways, replicating in their own fashion, privately."[536] The mitochondria are rod-shaped and look very much like *Paracoccus dentrificans*, a free-living bacterium that we still dignify with a scientific name of its own.[288] And they certainly contain their own DNA, which is self-replicating and quite different from ours. They have their own, rival hereditary system, a separate genotype that may make occasional contact with ours in the nucleus, but operates largely on its own.[203]

The usual way of looking at them is as enslaved creatures, captured to help us out with the business of respiration, just as chloroplasts help plants with photosynthesis.[463] Lewis Thomas notes, "It is a good thing for this entire enterprise that mitochondria and chloroplasts have remained small, conservative and stable, since these two organelles are, in a fundamental sense, the most important living things on earth. Between them they produce the oxygen and arrange for its use. In effect, they run the place." And he adds, "It is no good standing on dignity in a situation like this, and better not to try. It is a mystery. There they are, moving about in my cytoplasm, breathing for my own flesh, but strangers."[536]

Nor are they the only ones. There are other organelles, such as the centrioles that help with cell division, and the basal bodies that lie near each cilium or flagellum, which may well also have their own separate genomes.[408] And we wouldn't even be aware of them, couldn't see them at all, were it not for the fact that we owe several of our sense systems completely to the talents of yet another lodger.

Evidence for this comes from the slipper-shaped *Paramecium* that spirals through ponds by wavebeats of its exquisite array of cilia. If one of these creatures comes up against an obstacle, or some sadistic researcher prods it sharply at the front end, it stops and puts its eyelashes into reverse. This response is made possible by membranes around each cilium that respond chemically to stimuli by taking up or giving off calcium.[140] We now know that the same chemistry works in most of our sense organs, and electron microscope studies have shown that the light-sensitive rod cells in the retina of the eye, the chemical-sensitive hairs in the membrane of the nose, vibration-sensitive lashes in the fluid of the inner ear, and even the tail of every sperm cell all come from the same source as *Paramecium*'s cilia.[139] All have a characteristic structure of eleven fine longitudinal tubules arranged in a circular ring of nine around a central pair. And this pattern is identical to that of flexible spirochete bacteria which can still be found in a free-living state. The implication is that these microbes, or their ancestors, were among those that took up symbiotic residence in the larger cells of one of our ancestors, and that they still play relatively independent roles in the community of each one of our bodies.

I don't think there is any need to belabor this point any further. The evidence is in, and it is strong. Not only have we descended from single cells without nuclei, but we have brought some of them along with us. Or perhaps it is they who have been bringing us to this vantage point. Their presence within us, and their direct and close relationship to other similar cells outside us, raises whole questions of identity and matters of honor and dignity. In one sense, because we share the same invaders with all other men and every other living creature after its kind, we can feel a new affinity with life everywhere. We have close relatives, once removed, all over the place.

But there is another side to it.

We are beginning to understand how living things are organized by sets of instructions contained in the genes, those first replicators, which live on in a variety of survival machines. We know already a great deal about the specific way in which our personal appearance and behavior are governed by the genotype "me" in the nuclei of each individual. This unique collection of information can be thought of in terms of the "self" and could be where our awareness and consciousness originate. But now we are, through the actions of this very center, becoming aware of another independent system, of a genotype that has close connections with other forces outside our individual boundaries and owes us no loyalties at all. Mitochondria seem to be able to exist, in the form of free-living bacteria, without our help. But without them, we die in a matter of seconds.

There is, in the very fabric of our communal organization, in the symbiotic relationship of several more or less independent systems that go to make up our complex organisms, an essential imbalance. We are tugged two ways by the Lifetide. Ultimately we are highly sophisticated ecosystems in a delicate state of oceanic equilibrium. But at any moment in time, in the life and awareness of any one particular individual, there are tensions which are the inevitable expression of what seems to be a great divide. It may be that, given another billion years of evolution, these differences will be reconciled. Or it may be that they are essential environmental selection pressures, without which all life on earth would have become one vast, homogeneous, mindless gunk, humming happily to itself and covering the entire surface of the planet.

If that is so, *vive la différence*, but let us recognize that the tension exists and make it work for rather than against us.

I believe that the divide in our spirit is real, natural, and, given the circumstances, inevitable. Knowing what we now do about the possible origin of life in space, about the way in which the raw materials have been molded by the tissues of earth itself, and about the origin of complex life forms from disparate sources, we have what we need to help us understand how we and things here work.

I intend in the next section to show how we can use this information to explore identity, mind, consciousness, and the powerful tidal pools of the unconscious.

# PART TWO

# Confirming the Creator

By virtue of his reflective faculties, man is raised out of animal world, and by his mind he demonstrates that nature has put a high premium precisely upon the development of consciousness. Through consciousness he takes possession of nature by recognizing the existence of the world and thus, as it were, confirming the Creator. The world becomes the phenomenal world, for without conscious reflection it would not be.

If the Creator were conscious of himself, he would not need conscious creatures.

—CARL GUSTAV JUNG,
*Memories, Dreams, Reflections*

One of the most persistent stories in the folklore of science is that of the biologist and the flea.

This particular researcher is studying the behavior of a large and unusually cooperative flea. By virtue of years of practice and self-denial, and exercise of the precepts of classical conditioning, he has trained the flea to jump over a matchbox on command, whenever he yells the word "jump." Being a true scientist, he wonders what it is that gives the flea the ability to respond in this way. And he decides it must be the first of its three pairs of legs. To test his theory, he tears off the front legs and again gives the command to jump. The flea jumps. Revising his hypothesis, he then tears off the two middle legs. He yells, "Jump!" and the flea once more successfully clears the barrier. Finally, in a fervor of experimental zeal, he pulls off the last pair of legs, and once more issues the command. This time the flea remains immobile and refuses to jump the matchbox. So the researcher draws the inevitable conclusion that pulling off the flea's hind legs has made the animal deaf.[142]

Leaving aside the danger of drawing faulty conclusions from experimental data, and the ethical questions raised by the use and treatment of laboratory animals, this story still illustrates a most fundamental biological dilemma.

In order to investigate living things, we all too often tend to end up killing them. With the assistance of electron microscopy, we now have records of practically every single biochemical reaction that takes place in a living cell, and yet still know next to nothing about the secret of life in the combination of those parts. In fact, a recent paper in a journal devoted to perceptual psychology suggests that much of what we "see" with the aid of the electron microscope is illusory anyway.[252] Certain well-known substructures of the cell—such as the unit membrane, the endoplasmic reticulum, and the nuclear pores, which appear in such satisfying detail in the illustrations of most textbooks on cytology—may be merely artifacts produced by heavy metals used as stains in the preparation of specimens for the microscope.

Our stumbling attempts to investigate the nature of life begin to look rather like someone trying to describe electricity by determining the chemical and physical properties of the iron, copper, and rubber from which a dynamo is made.

We may never know the true secret of life—it may in the end prove impossible to discover and understand the relationship of life and mind—but we don't have to operate completely in the dark. There is no need to deduce the living properties of a cell, the results of having life, from its parts. Fortunately for us, we already know a little of what we are after. We can see the motor running.

The first to ask pertinent questions about life and to leave us a record of some tentative answers were the ancient Greeks. Despite the passage of two and a half millennia, some of the most significant literature on the matter is still to be found in writings that survive an intellectual chain reaction which began with Thales of Miletus and ended with the great Athenians, Plato and Aristotle.

It was Pythagorus, himself a student of Thales, who first introduced the important distinction between the particular and the general. He said, in effect, a triangle is a triangle is a triangle, and although a particular triangle may be in detail different from all others, there is a general quality about it which anyone can recognize as "triangular." Even a cat.[269] Somewhere in the mind of man, and possibly many other species, there is therefore an "ideal triangle" with which all other triangles are compared.

Plato, or perhaps Socrates, for it is often impossible to separate the two, elaborated this Pythagorean concept into his Theory of Ideas. *Idéa* is Greek for a visible shape or "form," and the theory suggests that every material object, anything from the Acropolis to a tennis ball, is no more than an imperfect and transient copy of an eternal pattern. Everything is merely an abstraction, a temporary shape or idea, produced from a more fundamental mold which constitutes the only true reality.

This is a concept which eerily anticipates conclusions only now being derived from discoveries in quantum physics, but for Plato it was merely a means to the elegant end of one of his dialogues. Since, he reasoned, we can experience the existence of an object by a number of different senses—we can both see and feel a pin which pricks us—some central system which coordinates experience must exist. For Plato this was the *nous* or mind, which was responsible for unifying sensation and relating it to the reality of ideas.

We now know, from research on the development of perception, that what an adult eye can see and understand depends very much on what that eye saw when its owner was a baby. In one elegant experiment, pairs of ten-week-old kittens that had been reared by their mothers in total darkness were put into a contraption in which one kitten could walk around freely while the other was restrained in a gondola that followed every movement of its partner. Both kittens saw the same things, but the one that experienced its environment in a passive way, like someone sitting at home watching television, proved later to have no valid picture of the external world and could not cope with reality until given its own chance to run about and explore things for itself.[248] Reality, it seems, is not self-evident. It is something we each have to learn for ourselves.

In several other experiments, kittens who wore goggles which prevented them from seeing lines of a particular orientation were never able as fully grown cats to see anything presented at that same angle.[247] And if only one eye is trained in this way, the other one takes over and compensates by seeing more of the blind spot and less of other stimuli which the spectacled eye can cope with on its own.[121] Experience is shared between the eyes to gain as much from the environment as possible. This suggests that, even at this simple physiological level, Plato was probably right. A pattern, an idea, is laid down in the brain, and what experience does is to provide the means for validating it. Our sense organs begin to look less like machines and more like theories about the nature of our environment.

A baby learns to see, it can't just start doing it. We learn by exploring things, by trial and error. Gradually we manage to decode the signals which reach us and interpret them in terms of real things. We learn to experience things directly, and to behave as though there was no need for any decoding, which means in the end that a bat using its sonar system "sees" obstacles acoustically as readily as other mammals experience them optically.

Knowledge is obviously not the same as sensation, but is the product of an act of integrating information acquired by the senses, of measuring it against previous experience, and making a reasoned judgment about it. Plato, as he always did, insisted that reason and judgment depend on argument, on dialogue; and that this dialectic could not take place between the mind and the body, but must all occur in the mind. Knowledge, he suggested, is the product of an argument between the mind and itself. And because it takes two to make a quarrel, or to have an argument, there must be two parts of the mind. He seems, in yet another inspiration of a kind that was common to early Greek philosophy, to have anticipated by more than two thousand years Freud's epoch-making recognition of the two minds—conscious and unconscious.

We now know for certain that the brain in a complex organism functions as an integrating whole, delegating responsibility for particular sensations to specific areas. There is a visual area in the posterior part of the cortex, there are auditory areas in the temporal lobes, and even a language center in the first and second convolutions on the left side of right-handed people. Every week our anatomical maps and our physiological knowledge become more detailed, but there is also an ever-increasing awareness that these mechanistic explorations are not enough. There is something else there that just can't be reached by scalpels or electrodes. Total blindness, deafness, or loss of speech can occur as a result of hysteria, or of direct suggestion under hypnosis.[58] And for these phenomena there seems to be no simple physical explanation.

If material reality is the product of one of Plato's "ideas," and these in turn are forms which can be controlled by our minds, wherever they may be, then we begin to be faced by the terrifying possibility that reality is actually created by the mind, and that we can change it simply by changing our minds.

The philosopher Karl Popper has constructed a three-level model of reality to include mind and its effects.[438] He proposes that the physical universe, which consists of beds and bodies and other obvious entities, exists in World One. World Two houses the mind in all its possible conscious and unconscious states. And in World Three are the contents of thought, all the products of the mind, including myths, legends, and theories, true or false. Many of the objects of World Three can obviously exist in material form as books or paintings in World One, but Popper is also suggesting that they have an independent reality.

Perhaps the best example is language, which can exist in a coded form in a World One materialization, or in electrical form in someone's brain as a World Two phenomenon. But it also exists in abstract, as something in its own right, in World Three. A man is World One, his mind is World Two, but his name can go on indefinitely in World Three long after the dissolution of the first two modes. It can even become a movement, a source of further activity and materialization in other men. Affecting reality in this way, it cannot be denied its own reality.

Richard Dawkins has coined the term "meme" to describe such a World Three object, a unit of cultural transmission.[125] Words, catch phrases, fashions, scientific theories, and ideas for making new and better mousetraps are all memes. And, as Dawkins points out, they are exactly like nonphysical genes, abstract DNA. Instead of leaping from body to body, they go from brain to brain, replicating by imitation. Once an idea catches on, it even begins to propagate itself almost without the help of Worlds One and Two. Memes are alive.

Nick Humphreys says that "memes should be regarded as living structures, not just metaphorically but technically. When you plant a fertile meme in my mind you literally parasitize my brain, turning it into a vehicle for the meme's propagation in just the way that a virus may parasitize the genetic mechanism of a host cell. And this isn't just a way of talking—the meme for, say, 'belief in life after death' is actually realized physically, millions of times over, as a structure in the nervous systems of individual men the world over."[125]

This is I think what Jung meant by "confirming the Creator."[301] God exists, if only in the form of an idea with high survival value, or the power to propagate itself, in the environment provided by our culture and our needs. Plato may or may not have one of his genes circulating in the gene pool today, but many of his memes still exist, kept alive by replication in a succession of gene machines. Ideas have a reality and an evolutionary momentum of their own. So in this second section, I want to look at the biological origins of consciousness, and set up alongside our growing knowledge of our physical roots an awareness of an ongoing psychical development which exercises equally powerful selective pressures. It is in the mixed medium of both processes that the Lifetide exists.

# CHAPTER FOUR

# FEELING: Sensitivity

One rainy day early in February 1951, an energetic young black woman, just thirty-one years old, reported to the out-patient clinic of the Johns Hopkins Hospital in Baltimore. Examination showed that she had a tiny purplish lesion on her cervix, and a sample was taken for biopsy. A few days later she was told that the growth, still less than an inch in diameter, was cancerous and ought to be treated. Cancer cells in the body are sometimes benign, forming no more than a local tumor, but those growing on Henrietta Lacks were malignant, rapidly detaching themselves to take root all over her body— and within eight months, despite everything the hospital could do, she was dead.[458]

Or at least most of her was. That tiny piece of her flesh taken for examination—rescued, as it were, before the first of her radium treatments—was passed by the resident gynecologist on to a laboratory run by George Gey, a pioneer in the field of tissue culture. This is essentially the art of convincing a small group of isolated cells that they are not confined to a glass tube in an incubator, but still part of a functional organ inside a warm living body, where it is safe to carry on reproducing. Gey was a skilled artist and grew the cervical cells in a roller tube of his own invention. After several weeks he realized that he had something very special in his hands. The cells divided faster than any he had ever seen. With a generation time of only twenty-four hours, they could, given optimal culture conditions and unlimited room to expand, have covered the entire world in a few years. In fact they very nearly did.

To conceal the true identity of Henrietta Lacks, she was called Helen Lane. Her cells, known as HeLa in the shorthand of tissue culture, began almost immediately to play a major part in medical research. Experiments can be done more easily, more quickly, on tissues than on intact human beings. And with a really reliable cell line like HeLa, which proved to be an ideal host for polio virus, a successful vaccine was developed in less than a year, and produced in quantity by 1954. Polio virtually disappeared, but HeLa was only just beginning.

Very soon it became the standard human cell line, used in millions of experiments in thousands of laboratories all over the world.

HeLa continued to grow rapidly and willingly for twenty years and everyone was terribly grateful, until a disturbing discovery was made in 1974. A team of Russians sent their American colleagues a set of assorted human tissue cultures, thought to contain something that everyone in cancer research was looking for, a human cancer virus. It turned out to be something else; but what was even more disturbing was that all the tissues, supposedly taken from a variety of Russian donors, were clearly recognizable pure cultures of indefatigable Helen Lane.

A Californian geneticist, Walter Nelson-Rees, began to check up on the pedigree of other cell lines and found that many of these were not what they seemed to be either.[403] Soon it was discovered that more than half of the established human cultures in prestigious tissue depositories and libraries everywhere looked suspiciously like HeLa. Experimenters who had devoted years of research to what they thought were kidney cells from Kansas City or breast tumors from Turin were in fact all working with identical versions of those unstoppable cervical cells from Baltimore. A curious paranoia spread through laboratories all over the world as it was realized that even a single HeLa cell carelessly transferred on a glass pipette to a carefully labeled colony of some other kind would settle in and take over in a matter of days. And even labs that had never worked with HeLa couldn't be sure that she wasn't lurking there under an assumed name.

Today HeLa's distinctive genetic characteristics have been carefully defined and everyone keeps a very watchful eye out for infiltration and subversion.[402] The crisis is temporarily over, but "Helen Lane lives," and nobody yet knows what made her quite so aggressive.

Under certain laboratory conditions, any cell from a mouse, a man, or a midge can be persuaded to fuse with any other cell no matter how foreign. Cytoplasm flows from one to the other, the nuclei combine, and it becomes for a time a single cell, a hybrid with the intact gene pools from two separate sources. Cell fusion is now a vital tool, responsible for much of our modern knowledge of molecular genetics. Useful chimeras of this kind are generally brought about by the addition of a special chemical derived from the protein cover of the Sendai virus.[165] This, as it happens, is a parainfluenza virus which has very obscure origins, appearing suddenly in Japan. It in some way acts as a catalyst, provoking ill-matched cells into the unnatural act of combining with each other. But Helen Lane carries out cell fusion without any such encouragement. Could it be that the cancer which grew on the cervix of Henrietta Lacks was induced in some way by a virus, whose characteristics are now part of HeLa? It seems likely. And if that is so, it is probable that the virus was a new one, and even possible that it did not originate on earth.

Under normal conditions, cell fusion just doesn't happen. In any ounce of good soil there are a hundred million bacteria, thirty million protozoans, a million algae, and another hundred million fungi, all milling around at close quarters as though they were part of an organized tissue. The bacteria alone on one acre of land may add up to several tons in weight. It has even been suggested that the soil itself is a living organism, with humic acid forming the body fluid in which its cells are seated.[259] But despite this abundance and proximity, the cells of different species don't fuse with each other. There are all sorts of other interactions. They eat each other, collaborate, accommodate, exchange, and barter. But fuse? Never.

The existence of this complex ecology among a wide variety of simple unicellular organisms belonging to a number of species suggests that the cells must have some way of deciding just what constitutes a species. They must have some kind of recognition system which makes it possible to distinguish one from another, self from others, me from not-me. There must be, even among single cells, a rudiment of identity.

In all vertebrates there is a highly developed immune system in which each individual organism, each member of every species with a backbone, produces chemical substances in direct response to invasion by a foreign body. This has become common and public knowledge since heart transplant surgery made rejection a newsworthy issue. The reaction clearly involves specific recognition by the chemicals or white blood cells concerned, because next time a similar invasion takes place the antibodies present in the blood remember it and are ready to deal with the intruder in a rapid and summary fashion. But nothing like an antibody has ever been demonstrated in an invertebrate. Insects and molluscs approach each invasion as a new event in its own right, to be dealt with from scratch. Normally invasion is only recognized as such if it is made by another species, by a potential parasite. And until very recently it was assumed that any invertebrate would accept cells and tissue from any other member of its own species. But this is not true. There is self-recognition even among simple colonial organisms such as corals.[285]

The individual coral animal is usually a small sac-shaped creature carrying a number of tentacles with which it catches food. Polyps of this kind join together in colonies whose size depends on the particular species. The gorgonians, as their name suggests, look a little like those three frightful sisters, growing into tree forms with tentacles writhing in hairy confusion on the ends of their branches. The colonies grow slowly by budding off new individuals, each of which adds to the structure and is automatically accepted by its neighbors as part of the system. And if such a growing colony is divided into two parts, these will readily reunite if given the chance. There is a clear community identity, but this doesn't extend to other colonies, even those of the same species.

Jacques Théodor of the University of Paris has found that branches taken from two individual colonies of the gorgon *Eunicella stricta* will not fuse with each other, although grafts removed from one portion of a colony will readily "take" when attached somewhere else on the same community. The branches which are strangers carry on with their own growth, but when they come in contact with each other, produce a barrier of dead cells, a sort of scar tissue that holds them apart.[534]

This recognition apparently takes place at a cellular level, because when small samples of tissue are isolated from gorgon colonies, they still have the ability to distinguish their own kind from others. They fuse with friends, but refuse to have anything to do with strangers. In fact both cell groups in such a confrontation go into a decline and their tissues collapse, but Théodor has made the fascinating discovery that this mutual retreat only takes place if the samples are of the same size. If one is larger than the other, only the smaller one disintegrates. However, it is not killed by the larger one, which is completely passive; it simply commits suicide. "It is autodestruction due to lytic mechanisms entirely under the governance of the smaller partner. He is not thrown out, not outgamed, not outgunned; he simply chooses to bow out. It is not necessarily a comfort to know that such things go on in biology, but it is at least an agreeable surprise."[536]

This means that even at a cellular level there are ways in which contact, communication, and recognition can take place. But it also shows quite clearly that there are advantages in numbers.

Referring back to that ounce of soil with its hundreds of millions of cells, something interesting takes place there which never seems to happen in equally rich, but more homogenous and less structured aquatic environments. Soil has less constancy and more pattern than water. The size of the grains, the air cavities, the moisture content, and the distribution of edible organic material all differ from one region to the next, and even in the same region over a period of time. This variety produces a range of ecological niches and stimuli which act as incentives to evolutionary development. And the fastest and most obvious direction for adaptation lies in cooperation, in becoming multicellular.[67] Yet again we see earth itself imposing pattern, acting as a template for growth and change.

Surprising as it may seem, there are even social bacteria. Groups of these most simple of all truly living things sometimes get together and act together toward a common end. *Chondromyces aurantiacus* begins in a little lemon-shaped spore that blows in the wind to settle on suitably moist soil. There each spore germinates and, as it splits open, several thousand tiny rod-shaped bacteria "stream out like a flame from the mouth of a dragon."[65] They then begin to glide in some mysterious way, without the aid of any visible moving parts, into congregations that join with similar associations from other spores, all the individual bacteria feeding and multiplying as they go along, until they form a considerable mass which is visible to the naked eye and looks like .a colorless slime. The slime pours through the soil, spreading first in one direction, then another, sometimes fanning out as though in search of new food. Each bacterial rod in the mass tends to follow in the tracks of another, like foraging army ants in line, dragging themselves along trails of slime. And all this goes on until food runs short. Then a strange thing happens.

The slime suddenly stops its perambulation and begins to pile up on itself, rod upon rod, each bacterium secreting more slime, until the organisms themselves are perched on top of a slippery skyscraper reaching almost a millimeter up into the moist early morning air. That may not sound very impressive, until you remember that each bacterium is less than a thousandth of a millimeter long. Transfer those dimensions to our scale and the slime tower is the equivalent of a building more than a mile high, which is five times the height of New York's World Trade Center towers. And from this grand height, bacterial spores are released to start the slime cycle all over again somewhere else.[132]

No one has yet discovered how these bacteria communicate

and how joint decisions are made, but there is another organism not very much more complex, which is far better known and has provided biology with some of its most magical moments.

*Dictyostelium discoideum* is another slimy character, usually known as a slime mold, but it is actually an ameba. In general, amebae are independent and solitary, pushing out desultory pseudopodia in an apparently aimless fashion as they ooze about damp places. At first sight these social ones are very little different, creeping through liquid films on the forest floor, engulfing bacteria and dividing relentlessly, again and again. And as long as there is enough food around, they go on doing just that. But when threatened by a shortage of suitable bacteria, they react in a very specific and deliberate way.[66]

The first ameba to sense impending starvation has a little chemical fit. It actually shudders in horror and releases a short burst of a chemical that has been identified as cyclic adenosine monophosphate, mercifully abbreviated to CAMP.[324] This signal travels out in all directions and alerts any other ameba within hailing distance, which is a circle with a radius of about ten cell lengths from the one who started the panic. The effect on those that pick up the distress call is instantaneous. They relay it in an amplified form, sending out a burst of CAMP a full fifty seconds long. Then they start moving directly toward the source of the first signal, and ooze without interruption for one hundred seconds, during which time they close down completely and don't respond in any way to other distress calls. So each ameba, having picked up and passed on the signal, becomes impervious to the calls of other ameba farther out who are in turn responding to the relayed signals. The result is that every ameba starts moving toward its CAMP neighbor, forming aggregation streams which tend in the end to converge on the original signal source which becomes an overall rallying point, a CAMP site.[405]

Looked at from above, a randomly speckled field of feeding amebae, all more or less equally spaced out, turns in minutes into a pattern of feathery crystal shapes as the organisms all begin to stream in toward several centers of attraction. And finally, after two or three hours, the process ends with all amebae in the area gathered densely around a small number of original "founding fathers."

Now each aggregation congeals into a sort of sausage shape about two millimeters long that begins to perform like a single multicellular organism. This composite creature, delightfully known as a grex, turns out to have sensations all its own.

Unlike its components, the grex is sensitive to heat and light and, developing a distinct front and hind end, it glides off at the great rate of one millimeter per hour in search of a warm, bright place to breed. It will travel toward even an extremely dim light with unerring precision, and will spend as long as two weeks oozing along a temperature gradient which it can recognize even if the difference between its front and back ends is only 0.0005 degree Centigrade.

Finally the grex comes to rest in the spot of its choice, and there this sausage of cellular slime mold performs yet another extraordinary feat. The separate amebae sort themselves out into work groups with a variety of different jobs and between them they seal a small group of workers, destined to become spores, into a capsule and hoist this up into the air at the end of a long, thin, well-anchored stalk. A group of identical and isolated cells succeed in getting together, decide on a course of action, undergo differentiation in a well-orchestrated division of labor and dedicate their joint effort in an extraordinarily altruistic way to the promotion of survival in a chosen few of their number. And, as if all this were not enough, the final fruiting body adds to the heat and light sensitivity already shown by the grex by becoming sensitive to gas. The stalked structures arising from each migratory grex carry on a gassy conversation with each other, leaning and bending and spacing themselves out so that as the capsule at each tip bursts, releasing the fully developed spores, it interferes as little as possible with its neighbors and stands the best chance of effective dispersal.[64]

It isn't food which induces the slime-mold amebae to become social. They can feed just as effectively, perhaps more so, if they remain small and isolated. It was actually lack of food that set the chain of social reactions going; but it is interesting to know, thanks to a recent discovery, that many bacteria on which the amebae feed also produce the CAMP chemical.[423] So what the starving founding fathers yelled to start the stampede was "Food!" It is also fascinating to know that this same chemical acts as an intracellular messenger in all organisms, even man. It mediates between hormones arriving at the cell wall and enzymes that lie inside the cell. This means that many of our most complex activities, such as the signals which hold the communities of strangers in our cells together, working to a common end, originate in the same ancestral reaction. What William Wheeler describes as "a strong predilection for seeking out other organisms and either assimilating them or cooperating with them to form a more comprehensive and efficient individual."[582] The process goes back to the first simple recognition system between cells, to the first time a cell was able to say, "This is me, and that is my food," and then to use this response to build up to the powerful recognition, "This is me, and that is not me or my food, but my friend."

The recognition of enemies, the development of an immune system, came only much later.[89] Adaptive immunity, which is the ability to recognize foreignness and to react effectively against it by creating antidotes, is now a vital part of our biochemistry. It protects us against infection by hostile invaders, but it also has the unfortunate effect of setting us against our friends. We reject in a paranoid fashion even kidneys donated with the best possible intentions. We have lost control of the system and with it an ability to discriminate for ourselves. We have even, heaven help us, become immune to ourselves.

Natural resistance to infection involves the production in the body of specific chemical antidotes, the antibodies, which combine chemically with the invader and put it out of action. Each antibody is tailor-made to deal with a particular foreign substance, its antigen. But the reaction between antigen and antibody not only kills an invading microbe, it also does some damage to the body itself. Many of the symptoms of infection are due not to the invader, but to our immune response. Generally this is considered a fair price to pay for overcoming the infection. You don't sue the firemen if they break down a few doors in the process of saving your house.

But the immune system is rather indiscriminate and reacts equally strongly to even harmless invaders. When golden pollen grains, too tiny to be seen, get into the covering of our eyes or into the mucous membrane in our noses, or even down into the lungs, all hell breaks loose. The immune system goes after the grains as if they were alive. But since they're not in any sense that affects us, and can't be killed or moved, they stick there, while the body blindly throws wave after wave of antibodies and white blood cells at them, like an army commanded by psychotic generals, hurling themselves at granite boulders in a field that will neither die nor give way in the face of even the most sustained attack. The whole area around the pollen grains becomes a battlefield, and normal tissues are injured and inflamed. Everything swells and reddens, and we cannot breathe or see normally. We cough and sneeze, our eyes itch, and our noses run; we end up feeling terrible until the pollen is finally washed away.[206] And it will all happen again next time that same pollen intrudes, because there are now antibodies to it built into the system, and hay-fever sufferers react to innocent plant spores in the same way that healthy people do to harmful bacteria, paying the same price to no purpose.

Now comes news that things have got so far out of hand that the immune system sometimes even goes into action without any kind of intrusion at all. It can turn the body's defenses against itself. This kind of unintentional suicide is called autoimmunity.

It can begin with some of our body cells undergoing a change induced by one of the new chemicals with which we have decorated our environment—by a poison on the food we eat, by a dye in an item of clothing, by a preservative or a pesticide with which we just happen to have come into momentary contact. Each is sufficient to spark off the auto-immune response, sending our defenses into action against some part of our own substance which just happens to have been labeled incorrectly. And even though we now recognize that this takes place, medical science is powerless to save the life of anyone so afflicted, because we simply cannot turn off the relentless conveyor belt involved in antibody production.

Auto-antibodies have now been isolated from a number of sufferers, but it is becoming increasingly clear in many cases that they are not the effects of the disease, but the cause of it. Pernicious anemia, which involves damage to the stomach lining; goiter, or malfunction of the thyroid gland; nephritis in the kidney; ulcerative colitis, damaging the intestine; rheumatoid arthritis; and perhaps even multiple sclerosis, which strikes out of nowhere at the brains of healthy young adults— all are probably autoimmune diseases. All examples of the body turning on itself for no known reason.

The normal immune reaction is rather like what happens in a beehive when a wasp intrudes. The worker bees on guard duty hurl themselves at the wasp and eventually sting it to death, but in the process it is likely that some of the defenders themselves get injured or even killed. As far as the hive as a whole is concerned, such losses are acceptable if the invasion is repelled. An allergy can in the same way be compared to a bee making the futile sacrifice of killing herself by stinging a petal that just happens to be blown in through the entrance to the hive. But autoimmunity resembles outright civil war, in which the bees end up stinging each other in growing hysteria until the whole community is totally destroyed.

This makes very little sense no matter whether one considers the species, the community, the individual, or even the genes themselves. Nobody benefits. And one can only come to biological terms with induced suicide of this kind by assuming that the society, the group of identities which have been subsumed in a complex organism, no longer remains viable. It is set to autodestruct when it can no longer function in the best interests of all those involved.

This is a melancholy notion. There may be something in it, even if only a moral, but my major concern at this moment is to explore the origin, and share with you what I believe to be the meaning of a great divide that exists in the spirit of nature. And to that end, only one aspect of immunity need concern us: namely, that it is a recent development in evolutionary terms, almost a refinement of social behavior, and that its origins lie somewhere in early sensitivity, in the recognition of one cell by another.

Early this century, there was a flourishing fishing industry in the Great Lakes of North America, based mainly on lake trout. Then in 1932 the Welland Ship Canal was completed, bypassing Niagara Falls, and opening the upper lakes for the first time, not only to shipping but to an invader from the sea. The eel-like lamprey *Petromyozon marinus* normally spends its larval life in freshwater streams and migrates as an adult to the sea, but faced with the availability of a whole new habitat, it reversed its movements and crawled up into lakes Erie, Huron, Michigan, and Superior. Adult lampreys or hagfish are all parasitic, attacking fish by attaching themselves with the aid of a sucker which surrounds their jawless mouths, and using a rasplike tongue to drill a hole in the host. Then they release a secretion which damages the tissue and breaks it down sufficiently to be sucked out along with the blood. The trout in the inland lakes had never been faced with a threat of this kind, and they quickly succumbed. By 1935 there were almost no trout left and the industry which relied on them had collapsed.[263]

Faced with the disappearance of their new supply of food, the number of lampreys in the lakes also dropped sharply as they returned to their old levels and to old associations with fish that had learned to live with them. There have always been trout in Lake Ontario and the rivers below Niagara, and many of them show scars where lampreys have been attached, but they survive because over millions of years they have developed a sort of immunity, a working relationship between host and parasite that allows both to continue in business— which is what the genes of each require.

Lampreys are the most primitive living animals with a backbone, and though they have no scales and no jaws or limbs, they seem to be direct ancestors of the fish on which they now prey. It has even been suggested that it was their predation which led to the development of the modern fish. All lampreys suck blood, and it seems likely that they always have; their structure is unsuited to living in any other way. But when they began, the only creatures with circulation on whom they could prey would have been others like themselves—a sort of suck-or-be-sucked situation, which would have been most unstable. Left to itself it would have ended, as things did in Lake Superior, with the death first of the host and then, by default, that of the parasite.

There were two ways evolution could move out of that impasse, and it seems to have tried both. One was for some types to develop skins tough enough to keep their parasitic relatives out, and we know this worked well because the fossil record from that period has a number of thick-skinned early vertebrates that look like armor-plated sturgeons. But the other, and ultimately more productive, solution was to keep the old soft skin, but develop ways of moving faster and keeping out of harm's way. This is what modern fish have done. But before the situation was resolved, there was presumably a long period during which it was a case of lamprey eat lamprey. And it seems to have been at this time, and in response to this direct pressure of a species on itself, that some of the suckers produced a chemical change in their bodies that rendered the secretions of other suckers harmless. They developed a resistance to their parasites, a kind of immunity. And today no lamprey parasitizes another, and nothing like an immune system has ever been seen in any organism more primitive than the lamprey.

We seem to have in this situation a special selection pressure and a specific response. The peculiar nature of the situation is that the tissues of host and parasite were so similar that conventional systems of defense were ineffective, which is exactly the problem we now face with cancer. Tumors are in effect parasites grown from the host's own tissue, parts of the body turned against itself by the introduction of a new set of instructions. And I find it highly significant that cancer too has never been seen in any organism lower down the evolutionary tree than the lamprey.[87]

I have already suggested that cancer may be caused by a new program imposed on an organism, either by an earth template containing contrary instructions, or by the arrival on earth of complete new codes evolved elsewhere. Now I would like to add to that the further speculation that such conflict of interests can only arise in communities of organisms that have acquired an identity which supersedes the collective interests of the separate parts. And I predict that one of the distinctive features of such superorganisms is that they will have immune systems.

In other words, I believe that the vertebrate animals, quite apart from having a backbone, which is after all only a structural convenience, have developed a degree of organization and complexity that makes them quite distinct from all other living things. They have crossed a threshold, made a qualitative leap, that leads ultimately to the evolution of true mind. And cancer is only one of the prices we have had to pay.

Our bodies are by no means static. A different person looks back at you each morning from the mirror. We produce a whole new skin surface once a week, the entire lining of the mouth is washed away and digested with every meal, each blink of an eye flushes hundreds of cells down the tear ducts. All in all we lose about a soup plate full of cells every day, and this loss has to be made good.

Fortunately it is, as bone marrow and generative tissue endlessly pump new cells into the system. There are millions made every day, and by any odds, perhaps even without new instructions from outside, there must be bad ones among them. It doesn't take much, a nucleus a little off-center or a protein missing one amino acid, and you have a faulty cell, a mutant, potential trouble. In the sea or in the soil, mistakes get washed away or buried. But in the body they must be destroyed. Our complex systems cannot tolerate any cell going off on its own, ignoring the group and getting on with its own replication, fulfilling its own deviant destiny.

The Australian Nobel Prize–winning physician Frank Macfarlane Burnet suggests that the immune system was developed specifically to cope with this recurring internal crisis, and that the fact that it deals also with intruders is merely a useful byproduct of the homicidal intensity with which it pursues homegrown criminals. He calls this watchdog activity "immune surveillance" and believes that cancer occurs only when it falls down on the job.[88] Others have criticized this notion on the ground that people tend to suffer from specific forms of cancer rather than having all kinds at once, which is what presumably would happen if there were general defects in the surveillance system. And still others, having found that tumors release antigens of their own into the blood, suggest that cancers grow because the immune system is tricked into wasting its energy on these decoys, making the fight in the blood-stream instead of in the tissues or the cancer cells themselves.[206]

The debates go on, but to me it seems clear that what we are dealing with in the immune reaction, whether it has to do with foreign intrusion or domestic mutation, is an identity crisis. And that is something which doesn't happen to invertebrates or flowering plants. They don't have personalities.

There is one phenomenon in the plant world which resembles the immune reaction. This is the device which many flowers use to avoid pollinating themselves.

"Nature," said Darwin, "abhors perpetual self-fertilization."[271] Apart from a few weeds that resort to sex with themselves in hard times, most flowers go to elaborate lengths to prevent pollen from the anthers from falling on their own stigmas. Some manage this by having the sex organs ripen at different times. The avocado even has flowers which open twice, once when the female organs are receptive, and once again when the pollen is ripe, all timed so that one tree's pollen can never reach its own flowers. There are actually two types of tree, one in which the males are ready in the morning and female readiness occurs in the afternoons, and the other in which the situation is reversed. Avocado farmers plant the two alternately.

This incest taboo exists for much the same reason as ours—mating with strangers is more likely to produce vigorous off-spring. And a few species of plant, such as lilies, petunias, and clovers, have, like us, a chemical incompatibility. In these, the pollen lands on the sticky stigma and begins to push its tube down the style, but very quickly recognizes this as one of its own and begins to withdraw and then to disintegrate. It is almost as though the plant was allergic to its own pollen, though no actual rejection takes place. It is the pollen which self-destructs, which commits suicide like the smaller of two clumps of coral cells, when it recognizes its mistake. Recognition in both cases takes place between proteins on the surfaces of the cells involved. There is a clear perception of the differences between "self" and "nonself," and an appropriate response to each. But this is an automatic coded response which does not imply true individuality.

There is now a huge body of literature on plant response.[544] The more academic contributors, the botanists, refer to "tropisms" or tendencies to grow toward light and away from gravity. Seen in speeded-up film sequences, bean shoots in a dark box actually seem to compete aggressively, lashing out at neighbors, reaching up and struggling with each other like serpents in a pit, straining to get to the light. But much movement of this kind is really irreversible growth in response to simple physical factors and cannot be equated with animal movements, most of which are reversible and can be repeated an indefinite number of times.

Other contributions to the literature talk about "emotion" and "feeling" and show quite clearly that there are specific electrical changes taking place in some plants in response to outside stimuli, such as sound vibration or the proximity of a complex radiating system like a human being. These reactions are vivid and often seem to be directed, implying that the plants are capable of recognizing individuals of another species and sometimes even of making value judgments between them. I have been involved in a lot of this work myself and can't help being impressed by the specificity and strength of some reactions, but am still concerned about the intrusive role of human observers in all the experiments, and have yet to be convinced that even the most complex plant system can establish its own identity to the point of possessing a distinctive personality.[571]

Much work in this area still needs to be done, but I believe that we are justified at this time in making the distinction between vertebrates and all other living things, and in attributing true individuality to the animals with backbones alone.

I am slowly, step by step, trying to establish this evolutionary sequence: Life originates in the existence of interstellar organic compounds; these essential precursors are patterned, on this planet at least, by information inherent in the earth; by virtue of this they become replicating systems which occasionally make the kinds of mistakes which give rise to a variety of forms; some of which collaborate in symbiotic relationships to produce relatively closed communities or cells which exercise certain kinds of sensitivity, and offer all the members considerable benefits; many of these cooperatives grow into societies of cells that are even more beneficial; and a few of the most complex associations actually undergo a qualitative change as a result of their complexity, which gives them a unitary sense, a "self."

And I am suggesting that this acquired individuality depends on having a complex organization of a certain kind, and only that kind. But before we can take this assumption as read, I must consider those invertebrate communities which seem in some ways to be similar, even to the extent of having a personality of their own.

The colonization of separate cells was tried many times during the course of evolution, and a few of these attempts, each with its own characteristics, have survived. One of the most simple is a sponge, in which there is a marked division of labor.[488] Each sponge has two basic types of cells, one of which forms a cuplike sac and another, with long whips or flagellae, which lines that sac. Together they form a pumping and filtration unit that brings water in from the environment, filters food particles out of it, and sends the stream of used water back into the surroundings. There are several different levels of complexity and further differentiation into active and passive cells arranged in various tissues, but it is no simple matter to decide whether a sponge "person" is the individual cell or the colony. I suggest that it is probably neither. Cells are in continual motion within each colony, taking on different appearances which depend on their position at the time. And if an entire colony is pressed through a fine silk cloth to separate each cell from every other one, this gruel will in time get back together again into a form which is typical for that species. But the new arrangement is completely different from the old one, with different cells taking on each of the necessary functions.[240] So it is impossible to decide just what constitutes individuality for a sponge. Can one call a casual arrangement of cells an individual? Is the new arrangement achieved by reintegration the same individual as the one that was squeezed through the cloth? And where does that individuality reside in the soup of separate cells? As a memory perhaps? A sponge meme?

Julian Huxley summed it up best by concluding, "It is better to believe in the historical individuality of the cells and to wonder at the idea of the whole's form that can thus penetrate the substance and absorb the individualities of its parts, robbing them of all ancestral freedoms."[272]

I prefer to take that just one step further and to believe that the basic units of individuality are the ancestral entities which together constitute a cell, and to stress that communal activity and sensitivity of the parts in a cell, or of cells in a colony, does not constitute identity or make a colony a person.

The next step up in complexity from sponges are the coelenterates, which include anemones, corals, and jellyfish. In one well-known colonial form called *Tubularia*, the soft stem which connects the polyps undergoes a spontaneous pumping cycle produced by waves of electricity passing between the cells. The communal movements of the separate cells have been called "concerts" and seem to be coordinated by a "conductor" or pacemaker cell of uncertain identity.[292]

This elementary nerve system reaches its peak in the open-ocean siphonophore *Nanomia cara*, which may well be the most sophisticated collection of cells ever gathered into a colony. At the top of each colony sits an individual modified into a gas-filled float, which gives buoyancy to all the other members strung out behind it. These include cells which specialize in protection, food collection, digestion, nutrient transport, and reproduction. But most interesting are the specialists called nectophores, which act like little bellows, squirting jets of water that propel the entire assembly along. By altering the shape of their openings, they are able to alter the direction of the jets and send *Nanomia* darting about vigorously, moving at any angle in any plane, even executing stylish loops for no apparent reason.[394]

This coordination is achieved by fine nerves connecting the individual cells, and the complexity of the system raises a basic zoological dilemma. Is the colony a collection of individuals, or is it a complex organism in its own right? On what basis do we distinguish a specialized cell in a colony like *Nanomia* from the organ of a mammal, in which it serves the same function? At what point does a society become so nearly perfect that we no longer have any right to consider it as a society, but must allow it to have an individual identity? The problem becomes particularly acute in the world of social insects.

Many ants, termites, bees, and wasps are so intensely social that it is useful to think of them as diffuse organisms. A colony of the African army ant *Dorylus wilverthi*, for instance, is a creature weighing almost as much as a man, but having twenty million mouths spread out over several acres.[592] One of the first to think about insect societies in this way was the enigmatic and much-maligned South African naturalist Eugene Marais, who said, "The termitary is a separate and composite animal in exactly the same way that a man is a separate composite animal . . . In every termitary there is a brain, a stomach, a liver and sexual organs which ensure the propagation of the race. They have legs and arms for gathering food; they have a mouth . . . Only the power of locomotion is absent, but if natural selection continues to operate, the final result may be a termitary which moves slowly over the veld."[360]

Working alone on that veld, more than fifty years ago, Marais was at a loss to understand the subtle machinery that held the termite society together, but today we know that his "mysterious power" is, at least in part, a group of chemicals called pheromones. These are chemical messengers or hormones which are manufactured by specialized cells and are capable of regulating the activity of other cells in remote parts of the body. In the termitary they are produced largely by the queen and circulate through the colony, passing from termite to termite, regulating their abundance and their behavior as each goes about the body of the colony, circulating through the corridors like cells in the bloodstream of an organism.[308] Among social bees, food is shared to such an extent that we can think of the hive as having a communal stomach. Information is shared in a similar way by pheromones or through the famous "dances," and is accomplished so effectively that the community looks like a single creature with a central nervous system and far-flying sense organs. Intruders are recognized by the body and dealt with in the ruthless manner of an immune system. Temperature is regulated very precisely, being kept at a high level despite the fact that none of the components is in its own right warm-blooded. Reproduction is handled by a small group of cells or caste members who are in every way equivalent to our ovaries and testes. And there are other workers which perform equally discrete functions, making them analogous to our liver, muscles, and kidneys.

Looking at insect societies in this way, we are forced to accept individual social insects as cells in a superorganism. No termite or honeybee can in fact exist on its own. It is sterile and no more capable of reproducing itself than a single red blood cell. Isolated from the organism, it very soon dies. The Bee and the Termite don't exist at all as separate organisms in their own right. They are totally artificial human concepts and ought only to be considered in context, as integral parts of a larger whole. When one of the parts dies, it is of no more consequence to the hive than the packet of cells we lose each time we comb our hair.

There are ample grounds for considering the hive and the nest as organisms analogous to any of the complex vertebrates, but we stumble still on the question of identity. Let's assume that some disruptive influence, say a tornado, breaks up a hive without killing a single bee, simply spreading all the insects out over the countryside. What then can we say about the organism? The body has disappeared, it has been disintegrated, but is the organism dead? If all the insects die, you're on safe ground, but what can you say about the organism when the scattered components are assimilated into other hives? Where then does individuality reside?

I must, if this argument is going to be able to resist flying off into purely philosophical areas, assume that identity has a biological basis. But it already seems clear that it is not founded on very fundamental structures; it cannot be located at the level of molecular biology. If it was, then it would presumably be something common to all life. So it must appear late in evolution. What is the difference, then, the essential organizational distinction between a bee and a bee eater?

Possession of a backbone is only symptomatic of that divide. In purely biological terms the vertebrate animals, those belonging to the phylum Chordata, have the following characteristics which set them apart from all other groups of living things:

Their main nerves are tied up together into a tubular cord running the length of the body and encased by a spinal column. This cord at one end runs into a well-defined head, where it expands into a brain, which is enclosed by a bony protective case that is compartmented by the main sense organs of sight, sound, smell, and taste.

Their blood system for the first time is completely closed and divided into two main longitudinal vessels, the lower one running forward and the upper one running backward down the body. And the fluid in this system is maintained under pressure by a special muscle, the heart.

Their metabolic functions are controlled and housed in discrete organs, the most important innovation among these being a well-organized blood-cleansing and waste-disposal system, housed in a kidney.

The details are not really significant, but it is important to grasp that the vertebrate animals are special because they are closed systems, spatially organized in this clearly defined way. No longer does blood simply wash freely about in a body cavity transporting food and lubricating tissue. No longer is there a simple, rather patternless diffusion of nerve fibers. No longer are sense systems scattered almost randomly over the body surface, so that clams end up seeing with their feet. One can't cut a vertebrate up into two or three pieces and expect these to go back into business in their own right. Functions are organized in the animals with backbone into clearly defined parts of the body, each with a special, and fixed, spatial relationship to all of the others. Vertebrates, for the first time, have a very specific form.

There is nothing to suggest that any of the vertebrate ingredients is, as such, any better than equivalent ones used by invertebrates. Squid have nerves and well-developed eyes, every bit as good as our own. A bee's sense of touch and vibration enables it to appreciate the fine details of dance even in the darkness of the hive. A moth's ability to detect just one molecule of the scent broadcast by its mate makes even a bloodhound look as though it has a cold in the nose. Insect exoskeletons are as varied and as carefully structured as any of our endoskeletons. Worker ants carry garbage out of the nest as efficiently as any kidney filters out impurities. Queen substance circulates from worker to worker in a termitary as effectively as any hormone travels through our bloodsteam.

There is no doubting that complex colonies of single cells such as the siphonophores, or societies of multicells like termites, represent some of the highest levels of sophistication ever achieved in three billion years of evolution. In many ways they are more efficiently adapted to their environment than even the most complex vertebrate. But they are different. Were it not so, some way-considering ant would be writing about the Lifetide instead of this sluggard.

Comparisons between men and ants are meaningless, because we have adopted entirely different solutions to the problems posed by the same survival pressures. The Belgian poet Maurice Maeterlinck was even moved to doubt that insects belong to our world at all. He said that "they bring with them something that does not seem to belong to the customs, the morale, and psychology of our globe. One would say that they come from another planet, more monstrous, more dynamic, more insensate, more atrocious, more infernal than ours."[168]

The differences are that great, but there are sufficient biochemical similarities for us to be equally certain that both we and they are earth-born. The proper comparison is not between ants and man, but between individual ant as component of its nest and each human cell as part of its body. The social insect's answer to the planet's problems has been to delegate responsibility to mobile units, while we have kept our components under more careful control. They are diffuse, while we are compact. In termites and wasps, the shape of the society is manifest mainly in the structure of the nest, which is often elaborate and highly characteristic, but within those confines organization is much less marked. We tend to think about insect politics as being rigid and totalitarian, but compared to the ruthless rule imposed on the cells in our bodies, ants live like anarchists.

The difference between us therefore boils down to one of form. Ant society and human body are organizations of great sophistication and equivalent complexity, with many analogous structures and functions, but man, because he is a vertebrate, has had growing complexity imposed on a basic shape. One can think of the vertebrates as being equivalent to an ant colony pegged out on a board so that the components are forced to fulfill their separate functions in a definite and more compact spatial relationship. They wouldn't be able to do this without considerable modification of their existing arrangements, in the process of which they would probably lose all their strange strengths and become something very much like a vertebrate. But, given enough time and a slightly different sequence of evolutionary events, the outcome might have been, and still could be, very different.

This notion of form as an evolutionary determinant is not new. It has been considered in detail by the physician Stephen Black, who in turn derived it from those marvelous ancient Greeks.[58] Aristotle in the fourth century B.C. concluded that mind was a function of form or shape, a product of the anatomical and physiological complexity of matter, which in turn was determined by something less substantial, like Plato's *idéa*. In other words, substance is capable of receiving an idea and taking on form like a block of flint in the hands of an early stonemason. And this form could then be used to fulfill a function, used as a tool like a hand ax, to make something even more complex.

The facts of nuclear physics as we now understand them certainly support this thesis. Matter is now assumed to be a form of energy—energy which has taken on a form and become substantial. And now the new knowledge of molecular biology takes us the next necessary step, which shows that the biological outcome of matter is dictated directly by the shape or form of living things.

The real transition from nonliving to living matter was made with the impression on simple organic substances of an essential shape, that of the famous double helix of DNA. The in-form-ation (and here the word itself implies what is going on) contained in subtle variations of this shape is transformed from DNA in the nucleus to the cytoplasm of the cell by messenger molecules which remember the shape, and with it are able to recognize, by their shape, the right amino acids so that they can be assembled in the right order, in the right place at the right time, and produce the appropriate protein. And so it goes on right up the chain, with shape and form at every level controlling the sequence of events that determine the species and make it unique.

Form is the shaping force of life. All form or shape contains information, and the more complex the form, the more information it contains. What I am suggesting now is that inherent in certain very complex forms, namely those arrived at after the millions of years of natural selection that produced the vertebrates, is the information necessary to determine the uniqueness, not only of the species, but also of the individual. That in the very configuration of our molecules and cells is the essence of identity. And that this is something which is not actually in the substance of any of the cells, but is a quality superimposed on them by their arrangement in space.

The distinction between substances and form is as simple as the difference between the chemical composition of clay and the shapes into which that clay can be molded. The substance of life is a small set of amino acids, but the forms are infinite. And in paging through the catalogue of possibilities, evolution has in our case hit on a form which is qualitatively different from the others. It is relatively independent of the substance and has the capacity to be formative in its own right.

One of the refinements of the form is an elaborate self-recognition and other-rejection system. And one of the consequences of having the form at all, of being an individual person, could be cancer . . . but there are others.

# CHAPTER FIVE

# THINKING: Awareness

The number of atoms in the universe has been estimated as $10^{80}$. And the age of the universe in seconds is something rather less, like $10^{18}$.[164] Therefore the number of distinct events which can occur in our finite area is limited by time. And the number of configurations in which a system of atoms can exist is much greater than the number in which it does exist. It follows from this that it is highly unlikely, even impossible, that any two samples of matter should ever be the same. Which means that no two individual organisms, no two cells, and none of the entities in those cells can ever be in the same internal state. Individuality is inevitable. But identity is not.

On hot summer afternoons, when the air near the ground is moist and warm, it becomes unstable and starts to rise. As it rises, it expands and cools, and water vapor condenses on particles of salt and dust to form droplets that concentrate in a cloud. The droplets are not of uniform size, so any which happen to be larger than average fall faster, colliding and coalescing with others in their path, growing at an ever-increasing rate until, if the journey is long enough, they fall through the base of the cloud as rain.

This is a more or less random process which, if the day is warm enough, can become cyclic. Air begins to move upward faster in some areas than others, currents develop, and as each updraft feeds on the warm damp air in its immediate neighborhood, a competitive situation develops. Rival currents vie with and even suppress each other, with only the fittest ones surviving, to give rise to the biggest clouds. The clouds accumulate, demanding matter and energy for their maintenance, growing from cumulus to cumulus and maturing finally in fully grown cumulo-nimbus form as adult thunderstorms.

A thunderstorm is a living thing, an organism with a definite and easily recognizable form. It stands on a foot of cool, hard primary rain, with a heel of secondary drizzle and a toe of rolling scud and squall, feeling its way slowly forward step by step. Drafts of warm air rise all about its body, flaring out at its head in an anvil of ice and hail. This anatomy is characteristic and well defined; morphologically it is divided into what can almost be called functional parts. Behaviorally, it is unique in the nonliving world, defying the normal tendency of most such systems to slide into the inevitable equilibrium of an inert state. The thunderstorm goes on doing something, moving, exchanging material and information with its environment, even reproducing itself by giving rise to daughter cells near its edges as the mother cloud dissipates through maturity into decay.[421]

Since thunderstorms are incapable of sexual reproduction, natural selection can act on them only as individuals, not as a species. Evolution is confined therefore to whatever happens in the life of each isolated storm, and no change, no mutation, nothing learned, can be transmitted to succeeding generations. A thunderstorm today is exactly like a thunderstorm that raged around a ruling dinosaur, or one that buffeted the first amphibian as it clambered out onto the mud of an early estuary.

In the drier parts of Africa, where thunderstorms are rare, it is often possible to predict the occurrence of one long before it even begins to form. You simply have to keep an eye on the army ants and wait for them to begin transporting their larvae from the nest to a safer place on higher ground. The ants have acquired this subtle sensitivity as a result of selection operating over millions of years in favor of colonies that learned weather forecasting. In this respect the ant-nest organism is a distinct advancement on the thunderstorm organism, but there are many similarities.

Each has individuality in the sense that it is a unique pattern of matter and will defend its integrity, fighting other colonies, struggling against rival updrafts, resisting forces intent on tearing the fabric apart. The parts work together for the benefit of the whole, but neither storm nor nest can in any real sense be said to have an identity. Neither is necessarily any more than the sum of its parts, even if in the case of the ant congregation the mathematics involved are formidable.

The largest colony of social insects yet encountered is that of the African army ant *Dorylus wilverthi* with more than twenty million members.[593] But not even this can begin to compare with the complexity of the units in a vertebrate nervous system. There are an estimated $10^{15}$ individual ants of all kinds alive in the world today, which is a figure almost as large as the number of seconds which have elapsed since the big bang that started our cosmos going.

This is very impressive, but as Lord Samuel put it, "To

expect us to feel humble in the presence of astronomical dimensions merely because they are big, is a kind of cosmic snobbery . . . what is significant is mind."[194] And there are more than $10^{15}$ great organic molecules in that fellow traveler of the mind, a single human brain.

This is not to suggest that identity and personality are merely the products of sheer quality. Something has to hold the alarming number of parts together. Consider the fact that each cell in your brain—and there are ten billion of them—receives an average of ten thousand connections from other brain cells, and has its own molecular structure renewed completely at least another ten thousand times during its active life. And add to this the knowledge that your brain loses more than a thousand such cells each working day, wiping out more than a trillion cross linkages in every twenty-four hours. Yet, "despite that ceaseless change of detail in that vast population of elements, our basic patterns of behaviour, our memories, our sense of integral existence as an individual" remain.[578] We grow and change, but remain recognizably ourselves—which means, very simply, that a sense of identity, and individual personality, cannot depend only on the substance of the body. It must be integrated somehow, somewhere else. Something has been added to make us more than machines. We have become, as Kant observed, ends in ourselves.[304]

This brings our argument out onto shaky ground, surrounded on all sides by seemingly bottomless chasms. We are becoming involved in the body-mind problem which has already swallowed several schools of philosophy whole. It has spawned the doctrines of materialism, reductionism, indeterminism, parapsychicism, epiphenomenalism, and parallelism, to name but a few; and the literature on each is large enough to fill any normal library. I do not have the competence nor the courage to get involved in this debate and I do not intend trying even to summarize the arguments here, but I am concerned with the tendency of most of the natural sciences to take the easy way out of dualistic dilemmas and always choose the mechanistic route.

Even psychology, which literally means "the science of mind," has become increasingly "the science of behavior"—which is not at all the same thing. It bypasses philosophical problems about the nature of mind and downgrades thinking to the level of just another form of measurable behavior. Psychiatry, I believe, touches the truth more closely, and I intend to draw on its insights later. But at this stage I can only express my disquiet with the way in which my science, biology, altogether ignores mind as part of the problem of life.

Speaking purely as a naturalist, with some experience of the fabric of life, I cannot see how any explanation of the physical world can be valid if it regards identity and self-recognition, which are forerunners of the mind, as being merely epiphenomena—accidental outcomes of the mechanical working of the machineries of life. I believe, and will try to produce purely biological evidence to prove, that there must be at least two distinct processes, two worlds, two different descriptions of reality. One of them includes all physical states and objects, all matter and energy, all the manifestations of order, including every living and nonliving thing. And another involves experience, subjective knowledge, states of consciousness, and creative imagination. Exploration of the first order through mechanistic theory has resulted in the vast and important progress made by material science. But I cannot see how any science which deals with life can afford to ignore the second order altogether. The evidence from evolution alone shows that the mechanism of mutation on its own could never, even in four billion years, have produced a single gene, let alone a mass of memes.

I believe we live in a world of emergent novelty and that the new inventions of individual identity and mind cannot be reduced or explained by referring only to the preceding stages. There has been a quantum leap, a qualitative change, which can only be partly understood by traditional scientific techniques. It seems to be true and inevitable that the very nature and future of science is to be essentially incomplete. The system is open-ended.

This leaves us still in the air, blinking nervously at the chasms all around, but I think that stability and even a certain amount of security can be assured by casting out one more, carefully directed, mechanistic anchor. I feel that it is necessary and important to examine carefully the consequences of having the special vertebrate form which seems to have given us personality. I think it will help to know for certain that unique individual identity is lacking among plants and invertebrates and that it is indeed present in, and common to, fish, amphibians, reptiles, birds, and mammals. Exploration of overall pattern in this way is a fundamental biological tool, something every good biologist learns as part of an evolutionary approach; and I am certain that it has an even wider application. I have already said that we can't expect to explain things by simply referring back to preceding stages, but I do believe that it helps to understand where you are, if you can look back and see quite clearly how you came to be there.

So, what is it that makes a man different from an ant nest or a thunderstorm? I have suggested that it is a special form which results in individuality and the recognition of difference and identity, and that this grows into true personality. But that is already going too far, too fast. There is one basic question that needs first to be asked about this phenomenon of recognition. What is it for? A lot of things, a vast number of experimental changes and modifications, are made during the course of evolution, but very few of them survive unless they have some use, unless they have survival value for the genes which make the decisions about who stays on to play the game again. Is there survival value in being an individual and in recognizing others in the same way? I believe so.

Reacting again as a naturalist, as someone who is familiar with the neighborhood, I think we need to redraw some of our boundaries. I am aware of man's special strengths, and saddened by the way we too often abuse them, but I am much more excited by our similarities with other living things than I am by our differences. Man is in many ways very special, perhaps even in some unique, but I am not convinced that many of these differences are necessarily qualitative rather than quantitative. I think most of the obvious distinctions between man and other living things are exaggerated by the fact that we have favored one particular characteristic at the expense of many others. We have become supremely self-conscious.

But we ought not to make too much of a mystery of self-awareness. John Eccles has said, "We must recognise that we cannot define at all how we came to exist as experiencing selves with our ineluctable uniqueness."[143] I respect the learning of this great neurophysiologist and I like the way he expresses it, but I don't agree. Self-awareness is not a mysterious substance or an unknown kind of energy, though it may have roots we have not yet recognized. It is another step along the line of evolutionary development that gave vertebrates their unique form and their special identity. I see awareness as a pattern of events in the brain, a very intricate and superbly organized pattern, but nevertheless a form, an arrangement, that has its own special qualities. It is directly comparable in biological terms with the phenomenon of heredity, which I choose to see as a pattern of chemical events in the reproductive cells of the body. It is the patterns that matter, and not the events themselves.

Our individual uniqueness is a direct consequence of the mixing of genes that takes place through sexual reproduction. We can say that individuality is the result of a chemical pattern, and go on to add that awareness is in turn the result of an electrical pattern which takes place in a special kind of individual.

I think it inevitable that awareness must have evolved a number of times in different animal groups, and I suggest that several of these evolutionary experiments survive and live alongside us. The ones which come most readily to mind are the apes and the whales, who have patterns of brain cells in many ways like our own. But I don't think we can afford to dismiss other apparently less "brainy" vertebrates.

At Harvard, pigeons have been trained to look at photographs and to peck at one of a number of discs if there is a human being in any of the pictures. The people can be shown clothed or nude, young or old, black or white, in any posture, and still the pigeons seem to be able to pick them out. Even the most fragmentary, and presumably unfamiliar, aspects of the human, such as a hand or a foot or the back of a head in the distance, were sufficient for accurate identification. The stimuli used in the tests were so varied and so complex that no simple conditioned response can be responsible. The conclusion seems inevitable that even the domestic pigeon is capable of forming a broad and comprehensive concept, when placed in a situation that demands one.[250]

Awareness of others is, I believe, a direct and necessary prerequisite of self-awareness. It is one of the rudiments on which natural selection must have acted, along with some mechanism of cultural transmission, to produce true experiencing selves, such as ourselves. Why bother? Well, I think that the adaptive importance of awareness is that it permits discrimination, such as the pigeons showed, and choice. And in the words of an evolutionary biologist, "The importance of experiencing oneself, of being aware, is that it allows a far more effective way of adapting to new and varied situations."[569] This alone would be enough to ensure that, once awareness appeared anywhere in an animal line, it would be selected for, and would persist.

I suggest also that there has been, in the way of much evolution, a mixing and blending of capacities evolved in different lines, and that seeds of awareness may have been drawn from a variety of sources. The ethologist William Thorpe suggests that "there has been, in the course of evolution, a gradual integration of many different sensory systems, having many different centres of coordination; and these, if not amounting to separate minds, seem at least to be independent centres of sub-conscious activity."[537]

I intend therefore to look back down the line for a moment to see if any traces of our passing remain, to try to find evidence among fellow species that such rudiments of awareness can and do exist, and could have been brought together like the components of a cell, to serve a common end.

Starting from first principles, from individual cells, the first and still one of the greatest problems posed by a multicellular organism is that of cell differentiation. How do the diverse types of cells in the same organism, all containing exactly the same set of genetic instructions, come to be so different? By calling the society an organism, we imply that it is organized, that the cells are arranged into distinct groups. But how does this happen? How do cells recognize and group with their kind? What leads them to avoid, and to position themselves in a specific spatial relationship to, cells of other kinds? And how do all the recognizably different cells arise from the same fertilized egg?

Genetics has shown us that there is massive redundancy, that each cell expresses only a fraction of the genetic information it holds. It selects and uses only a few of its genes. So, in a developing frog embryo, one cell chooses the information relevant to retinas and becomes part of the eye, while another selects a different instruction and becomes part of the spinal column. And it seems that in some organisms these distinctions are determined very early. If one of the two cells resulting from the first division of a frog egg is killed with a hot needle, the remaining cell develops into only half a tadpole. The different groups of cells in the frog owe their origin to different parts of the original egg, which may look uniform, but is apparently already subdivided into a mosaic of differential areas.[498]

The situation is quite different in that other favorite embryological animal, the sea urchin. In this invertebrate a single cell taken from even an advanced embryo will grow on to produce a completely normal larva.[124] And in many plants the selected cell doesn't even have to come from an embryo. A single phloem cell from a carrot has been cultured in the laboratory and grown into an adult carrot plant, complete with leaves and flowers.[513]

From these experiments we learn two vital things about recognition. First is that even at a cellular level, there is a difference between vertebrates and invertebrates. Invertebrate cells are much more generalized, more capable of taking on a variety of functions and changing character to suit the circumstances. So identical eggs laid by a termite queen can, depending on their experience of chemicals in circulation in the nest, grow into workers, soldiers, reproductive males, or even other queens. For them, identity is a very flexible matter. But it is quite different for most, perhaps all, vertebrates. In the fertilized egg of a frog, the destiny is already decided, with different parts of the cell each taking on distinct functions of their own. Identity, it seems, is clearly determined even before the organism is complete. And from experiments in which single starter cells of this kind have had their nucleus, with all its DNA, removed, it is clear that the ultimate identity depends only partly on the nuclear genes. The cell "is poised in a state of dynamic equilibrium between nucleus and cytoplasm, in which signals from the cytoplasm are necessary for the maintenance of the nuclear gene expression."[391] In other words, who we are depends not only on our parental genes, but partly also on guidance from the host of strangers we carry around in each of our cells. We are surrounded by fairies, good and bad, long before the christening.

And the second important fact to emerge from embryology is that any cell expresses only a part of its inheritance. Character, and therefore recognition which depends on manifest characteristics, involves the selection of one, or at most a few, features from the multitude available, which in turn makes a differentiated cell capable only of a selective response. It is important that a complex organism should be able to choose which bits of a baffling superabundance of stimuli it is going to respond to. We can't cope with everything all at once. So our eyes appreciate only thirty percent of the range of light coming from the sun, and reject the rest. Our network of brain cells, charged and firing without conscious control, edits things for us. Only the most simple cells lack this sensory barrier. It is a rather mournful truth that only they really perceive the universe as it is. The rest of us are living an illusion.

Ants recognize other ants from their own colonies by pheromones, chemical badges which they all wear, and exchange like passwords or secret handshakes at every meeting. Recognition for them depends on a single sensory cue, and any foreign ant who wears or uses this will readily be admitted to the fold. Even beetles and wasps that have learned to don this simple disguise are able to make their homes in ant nests and enjoy all the benefits without contributing in any way.

The beetle *Atemeles pubicollis* lives during its larval stage in the nest of the European wood ant *Formica polyctena*. And despite the fact that it steals food and even eats the young of its host, it is treated with astonishing cordiality. The ants not only tolerate these parasitic invaders, but actually feed, groom, and rear the beetle larvae as if they were their own. All this is achieved by a chemical the beetles secrete from glands in their skin; a pheromone so specific, so good at saying exactly the right thing, that even a crude lump of filter paper soaked in the secretion and left outside the nest will be greeted by the ants as a prodigal child and carried into the home for adoption.[260]

Individual identity can guard against risks of this kind.

David Lack many years ago showed that the male European robin *Erithacus rubecula* would attack a little bunch of red feathers even more readily than it would react to a realistic model of a stuffed male which had everything a robin should have except for the famous red breast.[330] This would seem to be an example of an automatic response, as rigid as anything in the careful program of an ant; but recent work on a robin's response to live rivals shows that it is much more adaptable. A territory-owning male robin will often recognize an intruder as a male by the detail of his song.

A robin's song is an elaborate succession of complex sounds, the exact sequence and phrasing, and even the tone, depending on the individual singer. It is normally a proclamation announcing the occupation of a territory and the readiness of the singer to defend his right to be there. If, when a robin first establishes itself on a breeding ground, one introduces another robin, or plays the recorded song of another male, the first robin has a violent reaction, threatening and posturing and singing his own praises. But very soon the song changes and the resident robin mimics the intruder, producing an exact copy of his song. This not only says, "Get out," but adds menacingly, "I'm talking to you, mister."[90]

In some birds, song patterns are fixed. A song sparrow *Melospiza melodia* which is reared from the egg by canaries and allowed to hear nothing but canary song still produces its own normal sparrow sound.[366] Cuckoos, of course, need similar programmed song patterns, because even under normal circumstances they never get to hear their true parents sing. Many other species of birds learn their songs by imitation. A young bullfinch *Pyrrula pyrrula* learns its song only from its own father, and if fostered by canaries will adopt the canary language and ignore the songs of other bullfinches, even if these are kept in the same room.[538] But the most common situation is for a bird to inherit the general form of its song, leaving each individual free to acquire the finer details for itself.

Usually this is done by imitation, which some species are extraordinarily good at. The mockingbird *Mimus polyglottos* of North America, the starling *Sturnus vulgaris* in Europe and the drongo *Dicrurus paradiseus* in India regularly imitate a whole range of other species in the wild. We don't know why this happens, but up to a point such mimicry may be useful. It is clearly important for the young of some species to learn the parents' individual calls, and to learn to recognize their mates' calls; but they must not imitate these too slavishly or they will never develop their own individuality. It seems that the tendency to copy and the ability to invent must be held in very delicate balance. This is how it probably is in nature for parrots and mynahs, who have very sophisticated vocal language for use among themselves. But in captivity the balance is destroyed and they end up imitating everything within hearing range, alive or not.[54]

Among white-crowned sparrows *Zonotrichia leucophrys*, which are abundant in thickets of the western United States, each geographical group has developed its own local dialect so that even two populations separated by a river have distinctly different accents.[535] And in small groups with their own distinctive sound, there is every chance for an individual to produce a unique pattern. The smallest of all groups is a pair, and in the dense forests along rivers in East Africa, pairs of boubou shrikes *Laniarius aethiopicus* produce the most vivid duets. The birds adopt phrases and rhythms which are characteristic of their locality, but each pair learn to sing together antiphonally, keeping contact by means of a liquid alternation of lovely bottle-round harmonic sounds which are their very own invention.[540]

There is obvious survival value in a breeding pair being able to keep together this way. Many species, such as swans, pair for life and are obviously able to tell their mates apart from other individuals. It seems likely that one of the factors selecting for the development of mutual recognition of this kind must have been the need to identify an old mate for new breeding. One study on the kittiwake *Rissa tridactyla* showed that a female who retained her mate from the previous breeding season bred earlier, laid more eggs, and had greater breeding success than other gulls that had to waste time and effort to know a new mate.[116]

The fullest development of individual recognition by voice occurs in seabirds which nest in large noisy colonies. Nobody who has ever seen a tern return to a dense colony and descend through a forest of sharp upraised bills to land next to the only bird in thousands that won't try to peck its eyes out can doubt that individual recognition has high survival value. What is surprising perhaps (until one remembers sea fogs and mist) is that, in the indescribable cacophony of a colony, this recognition should depend on sound rather than sight. The sandwich tern *Sterna sandviciensis* has a special "fish call" that a parent with food produces when returning to the nest, and spectrograms of individual callers have shown that each and every one has its own unique signature. Part of the cell says "sandwich tern," another part says "returning to nest with food," but both are prefaced by a distinctive curlicue which identifies the particular tern doing it.[270] The situation is similar in the common tern *Sterna hirundo*; it has been shown that four-day-old chicks respond immediately to a recording of their parent's returning call by cheeping, turning, and walking toward the loudspeaker. They are quite unresponsive to the playback of calls from any other member of their large colony.[512] And there is some evidence which suggests that some birds may even acquire this sensitivity by listening to the parents while they are still in the egg.

All in all, birds show a capacity for organizing sound-making and receiving with a precision that is seldom even approached by mammals. And much of this is dedicated to conveying specific information between individuals that clearly recognize each other as such. The same is true for reindeer, which can, it seems, recognize each other by voice. And it must be true of most dolphins and whales, which, like seagulls on a foggy day, have only their voices to keep them in touch. This may be more than enough.

Roger Payne's recordings of the humpback whale *Megaptera novaeangliae* off Bermuda show that it has a song which contains about two hundred notes, ranging from high-pitched squeaks to low and resonant rumbles, and which lasts for ten minutes or more. This is certainly a signal specific to the species and must be characteristic also of the individual singer, which can, under certain circumstances where there are temperature inversion layers in the ocean, communicate over distances of several thousand miles. The recordings include examples of one whale picking up on the song of another and returning it note-perfect, but with embellishments of rhythm and cadence that suggest individual creative activity. It has even been suggested that the acoustic memory of whales is so elaborate that they could perhaps listen to an entire symphony just once and then have a mental playback later remembering not only every phrase, but also the way they felt when hearing it for the first time.[392]

There are examples, even among mammals, of groups that have only insect-type cohesion in which the members recognize each other by a badge such as a smell, but are otherwise anonymous. Some social mice have a collective odor which is maintained by mutual marking with urine. They do not know one another, so there is no rank order and little conflict. Mating takes place without rivalry. But if one mouse is removed and anointed with urine from a foreign group, it is attacked in the same way that bees will attack an intruder to their hive.[395]

A more typical vertebrate arrangement is, however, that of the individualized group in which animals clearly are known to each other and establish, usually as a result of occasional fighting, a social hierarchy.

The phenomenon of rank was first studied in chickens, and so the arrangement of individuals from most dominant down to most submissive is known as a peck order. In birds the sequence is usually very clearly defined and maintained with only occasional skirmishes as growing juveniles work their way up through the ranks, or as individuals benefit by becoming attached to a particular mate. A low-ranking jackdaw female at once advances in the hierarchy when she mates with a high-ranking male.[352] But among social mammals, and in particular the higher primates, there is a confused network of interrelationships, which involve alliances and triadic associations that make for very complex politics.[312] This is a sophisticated system which relies on an individual's ability not only to recognize others, but to predict to a certain extent what effect his actions with or against others are likely to have on his own ultimate status. And party politics of this kind require not only an ability for abstraction, but very strongly imply a degree of self-awareness. This is a long way from the purely mechanical ability of mountain sheep to assess the strength of a stranger merely by the size of his horns and to admit him into his proper place in the hierarchy on this basis alone.[198]

From damselfish on a coral reef up to baboons on the savannah, there is clear evidence among all vertebrate groups of an ability to recognize each other as individuals, which leads to a unique and productive stability in social gatherings. At its most simple, this recognition may be a response only to certain characteristics of the mate such as the timbre of his voice. But at the other extreme, very sophisticated distinctions are drawn. Jane Goodall tells of a female chimpanzee in a large group who allowed all the males in the society to mate with her— except her two grown sons and her brother.[210] This not only suggests that our incest taboo might have a biological basis, but it shows a refinement of response that is highly self-conscious. To make social distinctions of that kind the chimp must have a very good knowledge of the identity of all group members and an added awareness of their long-term relationship to herself. She must be conscious of her existence as a self.

I have already implied that simple sensitivity, the reaction of a cell as a unit to its environment, is the earliest kind of feeling. During the long course of evolution, reactions of this kind were repeated countless times, and at some point which is still difficult to define, a change in organization took place allowing the creature to anticipate repetition of such episodes. This was the beginning of behavior, which has grown in many species to be highly complex, but which cannot be called awareness until it is further guided and developed by a finer feeling; by a sense of what is external and what is internal—not just the Me versus Not-me response of an allergic reaction, but a refined subjective sense of self-identity. We certainly have this, and there is growing evidence that we cannot deny such awareness also to many other vertebrates.[222]

Perhaps the most characteristic feature of this personal identity, the experiencing of self, is that it transcends space and time. It involves learning and memory, which bring together events normally remote from each other. In other words, it involves the creation of mental images or models of the environment which include the subject. This seems sensible, but it is extraordinarily difficult to prove. Our only way of establishing that mental processes exist at all in other individuals, even other members of our own species, is to examine what they can tell us about themselves. We can learn a great deal about their feelings and intentions by observing various forms of nonvocal behavior, but ultimately we depend on what they say they feel. This kind of interrogation has never been possible with other species—until now.

There are in virtually every culture myths and stories in the folklore that deal with talking animals and with devices like King Solomon's Ring that enable man to communicate with other species. These express our essential loneliness, our need to communicate, and our alienation from nature, but until a few years ago there seemed to be no way to bridge the man-animal divide. Then came a chimp called Washoe, and now it is almost necessary to divide our thinking about animal awareness into pre-Washoe and post-Washoe periods.

In June 1966, Allen and Beatrice Gardner acquired an infant female chimpanzee and began to teach her to "speak."[194] There had already been several attempts to teach chimps a vocal human language, and all had been monumental failures, because the ape's vocal apparatus and behavior are so different from our own. So the Gardners chose American Sign Language, or Ameslan, as their medium. Ameslan is a system of arbitrary visual signals made with the hands, each of which is analogous to a word in a spoken language. Washoe was kept on her own in an area in which no other means of communication was allowed. There were no other chimps present, and the humans who worked with her used only Ameslan among themselves and toward her. "Using a gestural rather than a spoken language was a stroke of brilliance that Washoe acknowledged by the rapidity with which she acquired signs . . . Five years later she knew one hundred and sixty words, which she used singly and in combinations in a variety of conversational situations."[348]

At the age of eighteen months, which is approximately when human infants begin to form two-word combinations, Washoe signed "gimme sweet" and "come open." And she demonstrated that the combinations were not being strung together at random, by using "open key food" to gain access to the refrigerator, or "open key clean" to get at the soap, and "open key blanket" in asking to have her bed brought out. By the time the first stage of the work with her came to an end when she was five, she had also developed the rudiments of syntax and was consistently ordering her words in the normal English way, by putting the verb between subject and object, as in "you tickle me."[195]

Almost immediately the scientific world reacted against the Gardners' claims that Washoe was using a language. One of the first criticisms came, surprisingly, from Jacob Bronowski, who denied that the gestures constituted a proper language on the grounds that Washoe neither asked questions nor was able to deny assertions put to her.[79] More recent work with her, and with other chimps, has shown that this is not true, and that it is difficult to see any real difference between a chimpanzee's use of gestural language and the use of ordinary speech by human children.

Washoe, now mature, has been "retired" to a colony at the Institute for Primate Studies in Oklahoma. She is, in effect, "an emissary from humanity, a Prometheus to chimpanzees who, it is hoped, will encourage a select group of chimpanzees to use Ameslan not only in communicating with people, but also in daily communication among themselves."[348] Both these are happening. Chimps in the colony, and in further isolation experiments, have begun to demonstrate remarkable inventiveness in the construction of new words and phrases. Washoe on first seeing a swan gestured "water bird." Lucy collectively groups all citrus as "smell fruits" and after tasting watermelon referred to it as "drink fruit," which is essentially the same word form as the English "watermelon." Her first painful experience with radishes led her to describe them forever afterward as "cry hurt food."[471]

But in our quest for evidence of early self-realization, perhaps the most impressive result of the Ameslan experiment has been the discovery that Washoe has learned to swear. Convinced of her humanity by having lived for five years with only humans as companions, she looked askance at other chimps on first contact with them, and described them as "black bugs." "Over a period of months, however, she began first to tolerate them and then genuinely to like them, although it is still an open question whether or not she categorizes them in the same class as humans," or as herself. And faced for the first time with the outrage of being caged in transit to her new home, she vehemently gestured "dirty Roger" and "dirty Jack," to erstwhile companions still free beyond the bars.[348] She clearly not only grasped the uses to which language can be put, but had very well-defined ideas about her own identity and status.

The use of that word "dirty" out of its usual context, but in an appropriate way, is fascinating. It is an example of the use of language as a tool. It involves memory in the transfer of a label from one situation to another, and it is in fact true symbolic behavior. The anthropologist Leslie White defines symbolizing (which he believes to be a uniquely human capacity) as an ability to freely bestow meaning upon a thing, and to grasp the meaning in a way that cannot be comprehended by the senses alone.[583] Washoe is doing precisely that, using "dirty" to describe a thing that is manifestly not unclean, and giving the word a new meaning. And she is also doing it in a way that carries with it the full weight of her personality, saying in effect, "Not only am I distressed and perturbed by my present situation, for which I somehow feel you are responsible, but I am sufficiently aware of my anger and its cause to choose to reproach you, in the hope that this might improve my lot or at least make me feel better." That is an unashamedly anthropomorphic interpretation of the gesture "dirty" as she used it in this situation, but I don't think there is any escaping the fact that Washoe was putting her feelings into words in a most creative way. She was saying exactly what she thought.

And in doing so she was demonstrating abstraction.

The brains of all vertebrates are similar in a general sort of way. They consist of three main parts. With increasing complexity along the line leading from fish through to mammals, the control center seems to move progressively forward, expanding as it goes until in man the forebrain is greatly enlarged. The hind and mid brains are certainly more primitive and seem to control reflex and automatic response, even in mammals. We are all governed to a certain extent by these old reptilian brain centers which are capable only of immediate and emotional responses. But the forebrain provides an alternative. It is capable of dealing with data displaced in time and space from the input of the moment. It is capable of holding and using memories, of making abstractions.

In man the balance has swung so far that the forebrain is dominant. It creates a model of reality abstracted from the mass of both current and old data, extrapolating even to include information yet to be received. It enables us to move freely in time and space, controlling our movement and our environment. It is the essence of mind, and was thought to be unique to man. But Washoe shows that it is not. We could see by surgery that the chimp had a cerebral cortex comparable to our own, but it wasn't until Washoe came along that we could be certain that it worked in the same way. It does. And now we are no longer alone.

Fascinating as it is, I don't want to get any further involved in the new research into animal language systems.[538] Symbolic communication has long been one of the main criteria used to establish man's uniqueness, and most of the work is dedicated to proving or disproving that ants, bees, pigeons, or monkeys can do it too. My reason for including any of the results here is simply to show that they also provide clues to the evolution of personality, and demonstrate some of the consequences of having an identity as an individual.

I suggest, in direct opposition to much linguistic theory, that it was not language that made man, but that human language, and other forms of symbolic activity, are direct and logical consequences of having that kind of organization which gives every individual a certain uniqueness and releases talents inherent in its form.

We come back again, almost inevitably, to the matter of form—or, to be more precise, to the form of matter; to shape and pattern as determinants of identity.

Most discussions about personality take the form of an argument between the effects of nature and nurture on an individual. Does one acquire characteristics by heredity from one's parents, or by experience from one's environment? The answer seems to be both, in differing degrees, depending on the species and its circumstances. But the friction between the opposing schools has obscured an important point, which is that heredity itself embodies an environmental factor—and provides further evidence of the far-reaching effects of form.

We get our inheritance from our parents, and for most purposes it is sufficient to assume that we get equal amounts from each. But that isn't actually true. From our mothers we get a large egg, and from our fathers a minute sperm. They don't share equally in providing our inheritance, even at this fundamental level, and there is considerable evidence to show that all the other differences between the sexes stem from this basic inequality.[422]

Before there were males and females—and some fungi still haven't discovered the difference—anybody could mate with anybody else. Sex cells were all the same, and each contributed the same amount of genetic material and the same quantity of food reserves to an embryo. But variability being what it is, there must occasionally have been some that were a little larger than others. These would have had an advantage in the sense that they would have got their embryos off to a better start, by giving them a larger food supply. So, over a long period, selection would have operated in favor of the big cells.

But the mere existence of large sex cells would have encouraged some small cells to take advantage of them. Any individual who made only small cells could cash in, and ensure that its genes were given the best opportunities with the least effort, by always mating with the larger ones. And it could achieve this end by making the smaller, cheaper cells more mobile, and capable of actively seeking the larger ones out. So eventually there would have been two types of sex cell with rival strategies. Richard Dawkins describes the large investors as "honest" and the small ones as "sneaky," and points out, "The sneaky ones would have evolved smaller and smaller size, and faster mobility. The honest ones would have evolved larger and larger size, to compensate for the ever-smaller investment contributed by the sneaky ones, and they became immobile because they would always be actively chased by the sneaky ones anyway."[125] The end result is that today one group of individuals has large sex cells, and we call them female; while another has small sex cells, and for convenience we call them male. And though both sexes live in the same habitats, and generally eat the same food and suffer from the same diseases, one takes a major role in the production and care of the young, while the other takes a more active role in courtship, is less discriminating in its choice of mates, more inclined toward promiscuity, and more contentious among others of its kind.[589]

A small difference in form, amounting only to a slight discrepancy in size, has led to some of the most dramatic distinctions ever to occur between individuals of the same species, carrying the same kinds of genes.

Genetic theory shows that on its long journey down through all the generations, an average gene can expect to spend approximately half its time sitting in male bodies and the other half in female bodies, alternately making small or large investments in the future.[492] Although the gene remains unchanged, its instructions differ radically. When riding a male gene machine, it says, "Body, mate with as many females as you can; don't bother too much about what they look like; and leave them as soon as you can be reasonably sure that the offspring will be cared for." When instructing a female body, the same gene says, "Be very careful whom you mate with; choose the most attractive male available, and try to find some way of committing him to you alone; delay committing yourself until you can be reasonably sure he will stay; and interfere with all his attempts to mate with other females."

These contrary programs provide a major difficulty for evolutionary theory, which cannot explain why a female should be less anxious to have her eggs fertilized than a male is to fertilize them—unless it is assumed that neither the species nor the individual matter, but that everything is designed to further the interests of the genes only.

Sociobiologists have now begun to make use of the mathematics of gene theory to produce models of such social situations in order to test behavioral strategies.[546] And they find that the typical male's interest in quantity, and the female's emphasis on quality, are exactly what each sex should do to ensure the maximum chance for the reproduction and survival of its genes. Only when viewed in this light do bizarre, apparently pointless, patterns of behavior, such as the tendency of the female praying mantis to eat her mate, make any sense. On the face of it, such lack of sensitivity does the male no good at all, but from the gene's point of view, a male mantis is investing in his children by helping feed the eggs that will be fertilized by his sperm.

In most vertebrates, sperm are relatively cheap to produce, while an egg is a valuable resource. And because of this discrepancy, a female need not be as sexually attractive as a male in order to ensure that her egg is fertilized. Her chances are good whatever she looks like, which means that it is the males who have to hustle and compete for attention. They have to resort to all sorts of advertisements to sell their wares and, as a result, they tend to be bigger, or more brightly colored, or more extravagant in behavior than the females.

There are some interesting exceptions to this rule.

In the forests of South America live shy compact chicken-sized birds with short, rounded wings. Male tinamou are the same size and the same drab gray-brown color as their mates. They do not build nests, nor do they indulge in any courtship behavior, nor put themselves out in any way to secure a mate or ensure fertilization. They break all the rules of the game, but they can afford to, because the male tinamou takes over most of the task of incubating the eggs and caring for the young. Parental investment is approximately equal, so they have come to look alike.

Among those strange fish called sea horses, the sex roles are totally reversed. It is the male who has a brood pouch and who becomes "pregnant," incubating the eggs within his body, giving birth on his own, and husbanding the young without aid from his mate. And it is significant, and a good test of the gene theory, that this reversal of roles brings with it also a reversal of strategies. It is the female sea horse who is more brightly colored and more aggressive.[589]

There is, however, one exception which seems to bend the rules without reason. In this species the female continues to incubate and to play the major role in rearing the young. She has a single egg which needs to be jealously guarded and wisely spent. The male produces millions of cheap sperm which he tends to distribute in a promiscuous fashion, often making little or no investment in a relationship. And yet it is the female who is brightly colored, who advertises her sexual attractiveness, while the males are predominantly drab. These exceptional animals are of course ourselves—modern men and women. And any extraterrestrial biologist looking at us could only assume that in our species the females compete for males.

Why? What does this mean? Well, with a species as complex as ours, surrounded as we are by cultural and racial influences, which often seem to have little or nothing to do with basic biology, it is risky to make generalizations. But the fact remains that we continue to be manufactured by our genes, and presumably their interest in their own survival hasn't changed. So either they have evolved a new strategy so subtle in its implications that its significance eludes even our best mathematical geneticists, or somehow the gene's control has been subverted or diluted.

I suspect the answer lies in both subversion and dilution.

Subversion involves the total undermining and overthrow of a regime. And in human terms this has been achieved several times by memes pitting themselves against the tyranny of the genes. For example, the custom of contraception is presumably not inherited. In the short term it works against the gene, whose interests lie in replication. But it could be argued that family planning and population control, by raising the standard of living, do in the long run act in the gene's interest. So perhaps a better example would be the habit of celibacy. This can't be passed on genetically. It is an evolutionary dead end, and yet it persists in our culture because it depends for its propagation only on an idea, which can be passed on without sexual reproduction.

Our strange inversion of brightly colored females and relatively drab males is, however, part and parcel of the reproductive process itself, and does not seem to be under memic control. If an idea were behind it, someone could presumably be found to tell us what the idea was. There are plenty of apologists for celibacy, but nobody seems to know why Western men traditionally consider it unmanly to be anything but drab. The answer, I believe, lies in dilution of the gene's influence, and I think we need to consider the possibility that there is now a rival program in action, and that it exists, also at an unconscious level, as an alternative to the nuclear control of DNA.

The gene's needs are best served by a strategy that involves sperm competing for the egg. That way both win. But the alternative program has disturbed this equilibrium, diluting the complete control of the genes by introducing a bias in favor of the egg. So, it is fair to assume, its interest must lie more in the egg than the sperm. Now sperm and egg are genetically identical—their nuclei carry the same number of chromosomes, the same quantity of genetic material—but they differ radically in cytoplasm. The sperm has virtually none, while the egg is rich in substance and nourishment. So the rival program must be concerned with, and perhaps located in, the extranuclear material of the egg.

The head of a sperm is almost pure nucleus or paternal gene material, though the tiny tail which moves it, and creeps along with it into an egg, is compounded of a centriole, a few mitochondria, and a propulsion unit that looks suspiciously like a mobile bacterium. The egg, on the other hand, consists of a small nucleus of maternal genes surrounded by the gigantic vitellus, which is a mass, thousands of times the size of the sperm, teeming with mitochondria, ribosomes, cytomembranes, Golgi components, and a whole host of other organelles that we know to be fellow travelers, cooperating in communal programs, but retaining to a very large degree their own separate identities. Many of them even go on speaking their own languages, carrying their own rival brands of DNA, each with its own genetic program.

I believe it is somewhere in this complex that we need to look for the new instruction.

According to the professor of ethology at Cambridge, "The unique instructions provided by DNA inheritance is at best but a partial explanation, and certainly not a sufficient explanation, of the experiencing self."[537] The gross structure and arrangement of an individual does seem to be the result of instructions carried by the nuclear genes, by our parental DNA, although part at least of the manner in which these orders are carried out depends on factors outside the nuclei. This much is known from studies of cytoplasmic inheritance and the role of chemical messengers in the cell.

Now I suggest that there are other, as yet largely unrecognized and often contrary, programs carried in the rival DNA of organelles—or perhaps not even coded in nucleic acids at all.

If they exist, these programs are probably not subject to the pressure of natural selection in the same way as the genes, and may not have changed in any appreciable way since they first joined the biological association in our bodies. It is even possible, now that viruses have come to look so much like mobile genes loose in the biosphere, that they come and go almost as they please. And even when they are with us, they're not all there, but just waiting, biding their time, dreaming their own dreams, becoming manifest in the phenotype and its behavior only quite recently in the course of evolution: when the forms capable of expressing them had been arrived at; when the ecology of communities became complex enough to give them proper expression.

I believe that these rival instructions are now part of our inheritance and are transmitted from one generation to another quite independently of the nuclear genes, possibly in the material of the egg. And that our reversal of sexual strategy is merely one small part of their actual and potential effect on our lives.

The fact that our females paint their faces and dress extravagantly may have given rise to billion-dollar businesses, but in the overall pattern of evolution, it is of little real consequence. As sexual roles change with the increasing emancipation of women, giving them the chance to be more aggressive in the choice of mate and opportunity, so the programs themselves have changed and relaxed their pressures on us. Men have begun to take on more parental responsibility and more colorful costumes, so that now we seem to have arrived at something like the unisex solution of the tinamou.

But the great divide remains. We are governed partly by nuclear DNA and partly by something else.

In the present state of our knowledge about this something else, it would be rash to assume that it is based on a unitary factor like a gene. It does in some ways seem more like a complex set of influences which function in a variety of different ways. It is not possible to be certain whether or not it has a physical identity or location at all, though I very strongly suspect that it is somehow closely connected with the organelles in our cells. But I think we can at least assume that there is in nature a force with an evolutionary momentum of its own, whose existence may have a physical basis, but whose expression is almost entirely psychical.

For the sake of convenience, I propose to give this system a name. It is by nature something peripheral and yet vital; something independent and yet producing dependence; something which is a constituent part of something larger, something forming an allotted share; something which seems to be there, but whose existence is by no means certain. There is only one word in the English language which says all these things. That word is "contingent." So I am going to call this alternative, nonnuclear, nongenetic set of influences the *contingent system* and refer to its functional units, if it has any, as *contingents*.

The Latin root of "contingent" is *contingere*, which means "to touch on all sides," and that too very aptly expresses the way I feel about this system, which is a little like air pressure, invisible and yet always there, flexing its muscles and pressing tirelessly on every part of the biosphere. I see it as something which has a collective rather than a personal nature, being part of a wider perspective, without the selfish interests of the genes, and yet as something which has at least the potential of exercising as direct and as powerful an influence over our lives as the nuclear government itself.

It is far too soon to begin building models of its mode of action, but I believe that the contingent system operates in the areas of decision where only a slight shift in equilibrium is sufficient to produce dramatic changes of direction. Like gravity, it is almost too weak to be measured and yet too powerful to be ignored. Contingents may affect genetic and embryological processes and be influential in this way in determining phenotype, but I suggest that their major influence over our lives lies in cooperation and competition with the nuclear gene system, and that it is this interaction which results in the development of mind.

Sensitivity in its most simple form arose as a result of normal Darwinian evolution, through natural selection and under the guidance and self-interest of the first true replicators, our nuclear genes. And I see no reason why such sensitivity should not blossom into awareness in precisely the same way. But I sense that the development of self-awareness, and its growth into full consciousness, requires the assumption of something else. And I suggest that it was a conflict of interests, a difference of opinion, between hereditary factors and contingent interests that provided that new impetus.

On second thought, a model might help to make this relationship clear. Think of our biosphere as a planet such as Earth. Now try to see nuclear genetics as a power like the sun, shaping life and holding it in orbit. Consider awareness in living things as an ocean on the surface of the planet, pushed about by the sun, but affected too by the lunar force of another power, the contingent system. The result is a conflict, but an orderly one, developing its own mood and rhythm, producing tidal effects. Sometimes the rival forces act in the same plane, producing sensory springs; sometimes they pull away at right angles in evolutionary neaps; but always the effects are there. It is in this dynamic situation that mind came to be, and it is to this combination of effects that we owe our existence as thinking, feeling, contradictory beings.

The tensions are eternal and essential. Without the tide we might exist, but we wouldn't know anything about it.

# CHAPTER SIX

# KNOWING: Consciousness

Some animals are definitely more equal than others. Just as we tend to divide the mammals up into man and the others, so the bird world can be split into two unequal categories: the bowerbirds and the rest.

The bowerbirds are confined to the Australasian islands, where they have developed patterns of behavior so advanced that they seem to have thrown even evolution into confusion.[205]

The impetus, as so often, was provided by sex. We have already seen how basic sexual inequality has resulted in an imbalance which forces the males of many species to go in for gaudy colors. In theory, the brighter the display, the more successful the advertisement is likely to be, with the sexual laurels going to the most extravagantly dressed males. In practice, however, there is a snag. Bright colors attract predators almost as well as they do sexual partners, which means that genes for bright colors tend to end up in someone else's stomach rather more often than genes for drab colors. So most males are forced to adopt some sort of compromise between their need to advertise and their need to survive.

In a few species, such as the peacock, they seem to throw caution to the wind and risk everything in one huge gamble. The peacock's tail is so outrageous that, despite the fact that it attracts carnivores, gets tangled in bushes and often allows him only a very short life, he is so attractive in those few halcyon days that he ends up fathering a great many children before he dies. And the genes like that.

But most species balance things out with greater moderation and confine their display to certain parts of the year, and often to selected areas during that breeding season. Natural selection seems also in some of these species to have decided that it is safer to keep gaudy males away from the nest, so they have been completely released from all nesting duties, leaving them free to gather in bachelor parties to disport themselves on special dance grounds or arenas. Most arena performers, such as the birds of paradise, clear the display area, carrying away sticks and pebbles to produce a more formal and easily recognizable site to which the females come for fertilization. And in this domestic activity seems to be the seed of an idea which started the bowerbirds off in their new direction.[367]

Somehow, somewhere along the line, one of their ancestors discovered that it was sexually just as effective to carry someone else's bright feather as it was to grow one of your own, and a great deal easier to deal with when predators were around. So began a line of evolution in which successive generations, and ultimately a number of separate species of male bowerbirds, came to look as drab and undistinguished as their mates. But they more than compensated for this discrepancy by going to the most extraordinary lengths to attract the attention of females in other, safer ways.

Some birds build bowers like thatched houses nine feet high, with several rooms, standing on circular lawns which are carefully manicured and decorated daily with bright-colored feathers and berries, iridescent insect skeletons, and fresh flowers. They tend these constructions assiduously, discarding and replacing the flowers when they fade, the fruit when it decays, and the feathers when they become bedraggled and discolored. The satin bowerbird *Ptilonorhynchus violaceous* even mixes up a pigment made from berries and charcoal and uses a piece of fibrous plant material as a paintbrush to decorate the thousands of sticks it builds into a ten-foot-long avenue across a forest clearing. Each bower is a stage set for the performance by the male builder of an intricate routine which is designed to end in mating, but the architectural, engineering, and artistic skill involved all add up to a pattern of behavior which "in its complexity and refinement is unique in the non-human part of the animal kingdom."[204]

There is a direct and inverse relationship between the complexity of the bower and the plumage of the builder: The more conspicuous his construction, the more drab he can personally afford to be. The forces of sexual selection, which are among the most powerful in nature, have been externalized. They have been transferred from bodily attributes to external objects. The bowers have become in effect external bundles of secondary sexual characteristics which are psychologically, but not physically, connected with the males, who come to look once more exactly like their women.

It is a lot easier to change your behavior than it is to alter your physical structure. This freedom from the constraints of the body and from morphological change has set evolution loose, and the evidence seems to show that the bowerbirds are evolving in all ways more quickly and more freely than any other family. Things are moving so fast for them that there are widespread examples of hybrids occurring not only between species, but even between totally separate genera.[204]

Donald Griffin of Rockefeller University says, "We have become so accustomed to concentrating on functional and adaptive aspects of these behavior patterns that we have neglected even to ask whether the animals have any awareness of the probable consequences of their behavior."[222] It is true that we do not as yet have any direct evidence to show that a bowerbird decorates its arena in anything more than a mechanical and compulsive way, but the results are so clearly aesthetic to our eyes that it is difficult not to feel that some real awareness exists. Very often the effects are achieved by careful, and apparently considered, selection and placement of a particular pebble or berry, when any bright object placed almost anywhere would have served equally well, if the only purpose of the display was to attract the unthinking attention of the female.

There are similar situations in the development and function of birdsong. The blackbird *Turdus merula* has a complex song sequence, and when one is singing well, by which I mean singing in a way which appeals aesthetically and musically to our ears, it seems to be exploiting this repertoire to the full. Such singing begins as a territorial proclamation, and when a neighboring male approaches too closely, the blackbird sings more vigorously in order to intimidate the intruder.[233] But he doesn't necessarily sing more musically. On the contrary, he seems to become upset and the song becomes loose and disjointed, pauses become longer, and phrases are left unfinished. So it looks as though the bird has to pay close attention to the form of its song in order to sing well by our standards. During the reproductive period, when the song is actively attracting mates and repelling intruders, it tends to be confined by these functional restrictions; but later in the season, when such reproductive restraints have been removed, it blossoms. Singing becomes more elaborate and more organized, in a manner so closely resembling our own ideas of musical form that it is difficult to deny that it is musically improved.[234]

These observations "amount to a quite serious argument for something like musical appreciation, albeit on an elementary scale, existing in a good many birds."[538] And for a kind of artistic appreciation among at least the bowerbirds. It begins to look as though it may be a serious limitation in our thinking to assume that such simple consciousness cannot exist outside our own species, and a bad mistake to underestimate its importance when it does.

The neurologist Roger Sperry says that "there seems to be good reason to regard the evolutionary debut of consciousness as very possibly the most critical step in the whole of evolution."[502] And the natural philosopher Karl Popper adds, "The emergence of consciousness in the animal kingdom is perhaps as great a mystery as the origin of life itself."[437]

Neither mystery may be capable of complete solution, but just as astrophysics at the telescopic end of the scale is providing insights into the origin of the universe, so molecular biology at the microscopic end is beginning to throw light on the nature of consciousness.

Probably the greatest breakthrough, and the biggest surprise, is a very recent discovery that even in the most complex organisms less than three percent of DNA in the cells is being used, and that the parts which are used, the instructions which are carried out, are taken in an apparently random way from a number of different points on the chromosomes. There is, it now seems, some sort of editor in action, who cuts up and subsequently pieces together little bits of genetic material from all over the place, like a radio producer intent on extracting a few minutes of concise statement from hours of recorded interview with a very voluble politician.

These findings have thrown biology into terrible confusion. John Rogers of the University of California says, "There is an air of bewilderment bordering on incredulity in molecular biology labs at the moment. It has even broken through into the research literature with the appearance of a paper whose title began 'An amazing sequence arrangement . . .' "[457] The problem is that genetics has always assumed that information is sequentially encoded in the DNA; and that this pattern is carefully transcribed onto a messenger molecule which carries it to areas where the code is translated into proteins. Nobel Prizes have been won for working out details in the sequence, but now it seems that it is not actually like that at all. DNA is not the Bible of life, not an encyclopedia of precise instruction, but is itself a flexible and dynamic system in which clusters of genes expand and contract, roving elements hop in and out, and each gene has to struggle for survival and expression in exactly the same way as the organism that carries it.[108]

And even that scene of molecular confusion gives no real notion of the complexity, because in some quite mysterious way, the offspring of an organism, which is judged by the expression of a minute fraction of its genes to be fit for survival, also inherit the enormous majority of genes that played no overt part in the proceedings. The problem is that we no longer have any way of judging exactly who is doing what. The phenotype, the external form of a living thing, can be as simple as a white blood cell and yet embody somewhere in its genotype all the complex history and potential of everything that ever lived. Max Perutz, the biochemist responsible for working out the molecular structure of hemoglobin, says of the new findings that "the gap separating physical structure and biological function is much wider than anyone expected."[430] The ghost in the machine looms larger than ever.

More and more it seems that form, the relationship of the parts to each other, is much more important than the parts themselves. Sperry says, "In searching brains for clues to the critical features that may be responsible [for consciousness], I have never myself been inclined to focus on the electrons, protons, or neutrons of the brain, or on its atoms . . . it has always seemed rather improbable that even a whole brain cell has what it takes to sense, to perceive, to feel, or to think on its own."[502]

Neurophysiology tends now to be more concerned with circuits of cells capable of perceiving pain or color, and with complex configurations that knit such circuits into a background of electrical activity. In these patterns on patterns it is possible to discern the effects of being conscious, of having a mind.

Wilder Penfield in his open-brain surgical work on epileptic patients found that one subject experienced an orchestral performance in great detail every time he stimulated one particular point in her brain with an electrode.[429] He felt at the time that he was somehow cutting in to a continuous videotape recording of the patient's emotional life, but nobody assumes any longer that memories are stored in any particular area in such a way. The brain is certainly not that simple. In Emerson Pugh's words, "If the human brain were so simple that we could understand it, we would be so simple that we couldn't."[444]

This is the Catch 22 of consciousness. Physiologists, psychologists, biologists, and philosophers have been trying for centuries to find a way around it, and have in the process produced thousands of books with billions of words, and one or two valuable insights, but we are still no nearer an answer. Two renowned scientists who have spent most of their lives researching consciousness were forced to conclude in the introduction to a recent joint work on the problem that "the link between brain structures and processes on the one hand, and mental dispositions and events on the other, is an exceedingly difficult one." It is "improbable that the problem will ever be solved, in the sense that we shall really understand this relation."[438]

So we are left with a phenomenon which everyone agrees is of vital importance, perhaps the most significant thing ever to have happened on earth, but which nobody really understands.

We have been conscious of the problem of consciousness almost since consciousness began, and each age has tended to try to define the question in terms of its own concerns. In the golden age of Athens, when man lived in freedom and slaves did all the work, consciousness was as free as that, a purely philosophical concern. But since Darwin and the Industrial Revolution, it has become more of a scientific question.

Could it be, suggest some physicists flushed with the success of quantum mechanics, that consciousness is a basic property of matter? It is possible that new knowledge of the relationship between particles will help us to understand the linkage between interacting minds, but unlikely that it will ever touch the true question, which is the awareness of such contact experienced by the minds involved.

Would it not be more reasonable, offer the vitalist biologists, to assume that consciousness is a fundamental property of protoplasm? It is an appealing idea that mind is common to all living things, but one difficult to investigate until we can teach an ameba the protozoan equivalent of the American Sign Language.

Perhaps, counter the behaviorists, consciousness doesn't exist at all? This revolutionary doctrine, which tried to reduce all conduct to a handful of reflexes, enjoyed some success in a world weary of war and longing for objectivity and stainless steel, but it was so patently ludicrous that it has now died a natural death. Nobody ever really believed that he himself was not conscious.

Which leaves us in the latter part of the twentieth century with the reasonable notion that consciousness began not with matter, nor with the origins of life, but at some midpoint in evolution. But there is still no agreement on where and when.

Current theories of consciousness assume either that it is in some way connected with learning, and ought to be isolated by experiments on the modification of behavior; or that it has its origins in some neural substrate such as the reticular activating system, and can be discovered surgically; or that it is something intangible, no more than heat given off by the wires, and therefore incapable of investigation; or that it is a metaphysical imposition, and therefore so remote that it would be, depending on your outlook, either unscientific or sacrilegious to pursue it at all.

I don't find any of these approaches very productive and tend to agree with Karl Popper that "What is . . ." questions lead only to arguments about the meaning of words and to discussion and definitions which are so abstract as to be useless. I think, and in this I am fully aware of exercising my own philosophical bias as an evolutionary biologist, it self-evident that consciousness exists in man and not in molecules. And I believe that it, or something sufficiently like it, exists in a number of other species—with a tendency for this number to grow every time some enlightened researcher gives his particular subject the chance to show what it can really do.

I hold therefore a belief in what has been called emergent evolution, which suggests that consciousness emerged at some point in evolution by the combination of a number of vital ingredients, all of which are important, but none of which on its own has characteristics of consciousness. Just as the property of wetness cannot be derived from the properties of hydrogen and oxygen alone, so consciousness differs in radical ways from its ingredients; but, like water, it would be impossible without its constituent parts.

This is in no way an explanation of consciousness, but it is at least a position from which it is possible to operate. All I am assuming is that consciousness is a fact, and that it is somehow dependent on the processes of life, and is not necessarily confined to man. This frees us from the circular argument about exactly what it is, and allows us to concentrate on what it does, and what it means to be conscious.

The principal effect of being conscious is that you have a strong and direct control over what happens. Bowerbirds have this to the extent that they set their own stages. It is pointless arguing about whether their ability is largely instinctive or partially learned, whether it shows simple sensitivity or an advanced form of awareness. What is important is to appreciate that possession of such sensitivity, whatever it is, has speeded up the evolutionary process among bowerbirds, and that consciousness, however we arrived at it, has given us similar adaptive advantages.

Functionally, consciousness is a feedback system. It lets you know how you are doing.

Feedback simply means taking some of the output from a system and returning it as part of the input. Any thermostat does this while regulating temperature. Biofeedback is the return of biological information to the person whose biology it is. Normally you can't see, hear, smell, taste, or feel your own blood pressure on the activity of muscle cells, but instruments can be designed to do this. And anyone who has access to such equipment can eventually learn to alter the frequency of his brainwaves, or to produce skin temperature gradients in which points on the hand just two inches apart register a difference of ten degrees, or even to stop the heart altogether for as much as thirty seconds.[82]

Decisions about when to alter blood pressure or bile flow or pulse rate, decisions about digestion or sweating or glandular secretion, are normally and most usefully made automatically by feedback circuits within the body. Interference with this system is likely to be beneficial only if the mechanism is faulty. But awareness of outside events—a feedback system which brings in news of things in the environment and of their direct relationship to the organism—must provide valuable intelligence. Simple systems of sensitivity in cells and organisms do just this, and when they get more complex, as in those animals that demonstrate advanced forms of awareness, they work even better. Any system which monitors itself and its environment in this way has already learned a vital trick. It has begun to simulate. It builds models of reality and runs rival strategies through the models, predicting outcomes and selecting those that on past experience promise to be the most rewarding, or the least punishing. No real intelligence is implied in such assessment. Computers do it all the time, and even flatworms can be trained to produce a particular response to a specific stimulus. This is all part of the gene's scheme for survival.

But real consciousness is one step ahead of such systems. It seems to arise when an organism's simulation of the world becomes so complex that it must include a model of itself in that world. This is the crucial step. This is the great leap forward. It means that for the first time the phenotype, the expression of the genes, the whole community of cells, has had a thought in its own right. It has done something which does not depend on the instructions of the genes. It has dared to stand up against the dictates of the government. This is the real revolution. This is what Descartes meant when he said, "*Cogito, ergo sum*"—I think, therefore I am.

In some ways the genes have only themselves to blame; they

have found selective advantage in developing survival machines with greater intelligence and great independence. But being themselves unthinking long-term creatures, they could never have anticipated the effects of advanced intelligence, which involves rapid and, from the gene's point of view, often unreasonable decisions. So, with the advent of consciousness, we have been thrown into what amounts almost to civil war.

Julian Jaynes of Princeton sees the conflict as a very recent development.[284] He uses laboratory studies on divided brains and a mass of archaeological evidence to argue that ancient peoples could not think as we do today, and were therefore not conscious. Being incapable of introspection, they were forced to listen to the instructions of the brain's right hemisphere, which they experienced as auditory hallucination, as the voices of the gods telling them what to do in circumstances of novelty and stress. He calls this condition the bicameral mind, and suggests that it was only as a result of worldwide catastrophe and cataclysm starting around the middle of the second millennium B.C. that the wall between the temples was breached, allowing the brain to talk to itself, giving rise to true self-reflection or consciousness.

During the past ten years there have been some fascinating discoveries about duplication in the brain which show that there are, in effect, two chambers in the mind. These began with the attempt of neurophysiologists to simplify research problems by splitting the brain down the middle before starting to study it. They found to their consternation that, far from halving their problems, they had actually doubled them, because the two cerebral hemispheres show an astonishing degree of independence.[501]

Roger Sperry and his associates at the California Institute of Technology have, in an attempt to control severe epilepsy in a number of patients, severed the corpus callosum, which is an isthmus of tissue which joins the two sides of the brain. The patients were relieved of most of the symptoms of epilepsy and in general appeared to function without any ill effects from the surgery. But in special tests it soon became apparent that the surgery had left each of these people with two separate minds, two distinct spheres of consciousness: one on the right side of the brain, connected only with the left eye, the left ear, the left hand, and the other crossing over in the other direction. So that if, for instance, a patient felt an invisible pencil with his right hand, he could describe it verbally with the speech center in the left side of his brain. But if he felt it with his left hand, he couldn't describe it at all. The right hemisphere knows no words. It was as though two independent people were involved.[197]

Another experiment tested reaction to visual stimuli by flashing the word HEART on a screen, with the HE to the left of the patient's nose and ART to the right. Normally if anyone was asked to report such an experience, he would say that he saw the word HEART, but the split-brain patients responded differently. If asked to name the word, they replied, "Art," which was the section of the word seen mainly by the right eye and projected to the left and vocal side of the brain. But when asked to point with the left hand to one of two cards which corresponded to their experience, they picked out "He," which is the projection received by the right, more spatial hemisphere of the brain. Similarly, when split-brain patients were asked to write and draw with either hand, they could write only with the right hand, and draw only with the left.[413]

These, and a multitude of more recent experiments, leave no doubt that each of us is normally in two minds about everything, and that there is a concentration of analytic, logical, verbal, intellectual, and essentially masculine activity in the left side of the brain, while the right is predominantly emotional, musical, spatial, intuitive, and feminine. In very simple terms the left brain is a doer and the right one a dreamer; and reaching for knowledge with the right hand results in science, while the left hand tends to come up with an artistic solution.[84] So, if you are a writer or a scientist, damage to the left hemisphere could prove disastrous; while if you happen to be a musician or an artist, a damaged right hemisphere will probably end your career.[413]

One of Roger Sperry's experiments is particularly revealing about our essential duality. In this, a red or green light was flashed to the patient's left eye, which sends news of the signal directly to the right brain. The patients were asked verbally to say what color the light was, but because speech begins in the left brain, which wasn't stimulated by the light, they denied seeing any flash at all. So they were asked instead to guess the color, the assumption being that they would guess correctly approximately half of the time, because the side of the brain which was doing the guessing was disconnected from the side which had seen the color and knew the answer. But all the patients scored far better than chance alone would allow.[500]

What happened was that the patients were "second-guessing." If they made an incorrect guess, they often frowned or shook their heads and then quickly corrected their verbal answer. The right brain had seen the light, then heard the left brain make the wrong answer. Having no access to words, it could only use another kind of signal, one available to body language and gesture, to let the left brain know that its answer was incorrect. So each answer was being provided in effect by two individuals with separate information and ability who reached consensus as a result of communication. This would be a normal response if in fact two people had been involved, but the whole dialogue took place within one split-brain patient. For those of us with an intact corpus callosum the dichotomy is not normally so dramatic, but we are all of us essentially schizophrenic.

All vertebrates have brains divided longitudinally by a deep cleft. This is partly a result of being bilaterally symmetrical, but space in the cranium is limited and one cannot help wondering why valuable volume should be wasted by duplicating all the control centers. The classic answer is that duplication provides a reserve in case of injury, but then why not also two hearts, two spleens, a second penis? That can't be the whole answer.

The only hard evidence we have for a breakdown of duplication is the lateral specialization in modern man. And this seems to have occurred mainly to cope with the huge load placed on the brain by the evolution of speech, but I think it is unrealistic to assume that this alone is responsible for consciousness or must necessarily be unique to man. Julian Jaynes suggests that language preceded consciousness, but this too is a construct so awkward that I must assume that whatever he believes began a mere three thousand years ago, it is not what a biologist normally understands as consciousness.

Humans are able to generate a rich variety of sound because we have a wider throat than our tongue-tied relatives. And the fossil records show that we have had this advantage for at least three million years. The initiation of movement in the face, lips, tongue, and larynx is controlled by a swelling on the lateral cortical lobe called Broca's area; and the coordinating of speech is somehow associated with another brain bump named after its discoverer, Wernicke.[338] Ralph Holloway of Columbia University has shown, by taking latex molds of the inside of the brain case, that these same formations are found in African fossils belonging not only to our direct ancestor *Homo habilis* but to several of our distant relatives, the Australopithecine apemen. "The evidence suggests that, contrary to what is widely believed, the human brain was not among the last human organs to evolve, but among the first. Neurologically speaking, brains whose organization was essentially human were already in existence some three million years ago."[261]

Which implies that language did not appear suddenly out of nowhere in the last fifty thousand years. It has its origins in biological machinery which has been accumulating for millions of years, and simply found its first concrete expression in those of our early ancestors who moved out of the forests and onto open terrain, where social food gathering and organized hunting provided a greater need to speak, and more to talk about. This is how evolution works. Washoe and the other apes now communicating with men didn't suddenly grow all the neural apparatus required to produce their talents. The equipment was there; all that was lacking was the incentive, the necessary environmental nudge.

What was it then which nudged us out of reflex reaction into considered response? What was the stimulus which led to thinking, awakening, being independent of the genes? I believe it was the process of feedback itself.

Try, as an experiment, to remember how many windows there are in your house. Most people, even those who say they have no visual imagery, can easily form and examine a mental picture like this and even count the windows in the mind. In psychological jargon, this is called an internal representation of sensory origin, but let's keep it simple and call it a model.[401] Simulation or model building of this kind certainly plays a major part in behavior which requires the recognition of predators, or specific individuals of the same species, or the memory of places and patterns which comes into action in orientation and migration. Birds do it, bees do it, every creature that has ever displayed any kind of learning probably does it to a certain extent. And all that is necessary to raise this ability, this kind of feedback activity, to the level of true consciousness is to add a direct awareness of this model in the mind.

Vertebrates have a very precise awareness of their surroundings. Their perception is acute. But, in the words of another definition, "Perceiving is a conscious state only if I am aware that I am perceiving."[77] The operative word here is "I," the essence of subjectivity—which means not only that the community of cells with their genetic jockeys must come to recognize their collective identity as an organism, but that they must also be able somehow to put at least a little distance between themselves and the image of themselves, and see it in isolation.

On one occasion the Gardners showed their chimpanzee her

image in a mirror and asked, in Ameslan, who the reflection was. She replied, "Me, Washoe."[338] Chimps clearly recognize themselves as discrete entities, and this cannot be done unless the animal is aware that such a thing as self exists. Perhaps this capacity to form that is in effect a mirror image in the mind has come about because there are two minds which can reflect, and reflect on, each other.

All the evidence points to the fact that visualizing or model-making, at least in man, takes place in the right hemisphere of the brain, which presumably means that reflection on this image occurs in the left. Duplication in the brain, which may be totally fortuitous or even encouraged by the genes as an emergency system, has provided the mechanism necessary for internal psychological feedback. It makes it possible for an organism to think about itself and to use this new awareness as part of the input on which future behavior may be based. It allows an organism to think for itself, and therefore to be.

As a biologist, and particularly as one privy to the results of the new work on our nearest relatives, I cannot see any justification whatsoever for assuming that only man has the capacity for recognition and self-awareness. And even if I were to concede that consciousness in all its full glory is manifest only in man, I must still insist that it did not spring fully armed and full-formed, uttering its Cartesian war cry, from the forehead of modern man. It arose gradually, in the normal evolutionary fashion, from rudiments already present, not only in ourselves but in many other species.

Jaynes may be right in assuming that the dichotomy between the two halves of the brain was responsible for giving rise to consciousness, but one has to look at his chronology as an allegorical construct, similar to the one used in biblical accounts of creation. Three thousand years of consciousness is no more realistic than six days of Genesis. Even our current estimate of three million years may be parsimonious. I am more inclined to put the first stirrings of self-awareness somewhere in the Paleocene, almost a hundred million years ago, when the dinosaurs were dying and the modern birds and mammals were coming to early fruition along with the flowering plants.

Conflict between the cerebral hemispheres is very marked in split-brain patients, but under normal circumstances we use both parts of our brain, the two sides of our personality, to complement each other. This is a useful division of labor, like the sharing out of sensation among a number of discrete sense organs, but it is important to appreciate that specialization in the brain has not gone as far as it has in the eye or ear. A man with only half a brain can, given time, develop all the normal faculties of logic and intuition, speech and imagery, on one side alone. Two brains are not essential for the development of a well-rounded personality in an individual. But for our species, for the current collection of human individuals, things would not be as they are if at some point we had not possessed the necessary duplication to come to terms with our selves.

So, while I acknowledge the fact of laterality and its effect on our consciousness, and recognize reflections of duality in literature and myth, I believe that it is wrong to reduce all human response to expressions of the relative dominance of the two hemispheres. There are conflicts between reason and passion, logic and intuition, mind and heart, men and women, and these may work themselves out in a kind of intercerebral rivalry. They may even result in neurotic or psychotic malfunctions, but I sense, perhaps mainly with my own right hemisphere, that the basic disturbance goes deeper than that. Why else should a man with only half a brain still show all those signs of conflict-in-resolution that we have come to recognize as normal individual personality? You can't change this by removing or substituting one lateral half of the brain for the other. The two are mirror images and, as far as form is concerned, identical. To change character you have to destroy the form, which means something like a prefrontal leukotomy which disconnects the front of the brain from the rest. And if the brain is complete, this means cutting the connections on *both* sides.

This might imply that the essential conflict is between the newer parts of the forebrain and the more primitive parts in the mid and hind brains—between the mammalian and reptilian memories. And in a sense this is probably correct, but I doubt that it is possible or even necessary to isolate the command centers of the opposing forces in any spatial location. The war is between the old selfish instructions and the new self-awareness, between genotype and aspects of the phenotype, between the needs of the replicators to keep on doing their thing, which is replicating, and the desire of the organism for identity. The battle lines are drawn between orders and ideas. Where the two coincide, a truce is declared and progress takes place by leaps and bounds. But where they disagree, skirmishes are fought in the no-man's-land of the mind, and ambivalent we, with all our special strengths and peculiar frailties, are the result.

I believe that the seeds of this conflict are sown in every cell by the presence there of nuclear DNA and factors connected with the contingent system, and that just as the presence and pattern of a certain number of cells behaving in a certain way can produce sensations such as sight or sound, so the mere existence of contingent factors in sufficient numbers in certain critical configurations could account for their recent intrusion in evolutionary affairs.

There is a biological analogy which makes this process clear.

The behavior of the Japanese monkey *Macaca fuscata* has been studied intensely for more than thirty years in a number of wild colonies.[277] One of these is isolated on the island of Koshima just off the east coast of Kyushu, and it was here in 1952 that man provided the monkeys with the right sort of evolutionary nudge. Provision stations were established at selected sites in the range of the troop. Normally young monkeys learn feeding habits from their mothers, who teach them by example what to eat and how to deal with it, and in these macaques the behavior had grown to a complex tradition involving the buds, fruits, leaves, shoots, and bark of well over a hundred species of plants. So they approached the new artificial food supplies equipped with a formidable array of behavioral predispositions, but nothing in their established repertoire enabled them to deal effectively with raw sweet potatoes covered with sand and grit.[310]

Then an eighteen-month-old female, a sort of monkey genius called Imo, solved the problem by carrying the potatoes down to a stream and washing them before feeding. In monkey terms this is a cultural revolution comparable almost to the invention of the wheel. It involves abstraction, the identification of concept, and deliberate manipulation of several parameters in the environment. And, reversing the normal trend, it was the juvenile Imo who taught the trick to her mother. She also taught it to her playmates, and they in their turn spread the news to their mothers. Slowly, step by step, the new culture spread through the colony, with each new conversion taking place in full view of the observers, who kept a constant watch right through all the daylight hours. By 1958, all the juveniles were washing dirty food, but the only adults over five years old to do so were the ones who learned by direct imitation from their children.[311]

Then something extraordinary took place. The details up to this point in the study are clear, but one has to gather the rest of the story from personal anecdotes and bits of folklore among primate researchers, because most of them are still not quite sure what happened. And those who do suspect the truth are reluctant to publish it for fear of ridicule. So I am forced to improvise the details, but as near as I can tell, this is what seems to have happened.

In the autumn of that year an unspecified number of monkeys on Koshima were washing sweet potatoes in the sea, because Imo had made the further discovery that salt water not only cleaned the food but gave it an interesting new flavor. Let us say, for argument's sake, that the number was ninety-nine and that at eleven o'clock on a Tuesday morning, one further convert was added to the fold in the usual way. But the addition of the hundredth monkey apparently carried the number across some sort of threshold, pushing it through a kind of critical mass, because by that evening almost everyone in the colony was doing it. Not only that, but the habit seems to have jumped natural barriers and to have appeared spontaneously, like glycerine crystals in sealed laboratory jars, in colonies on other islands and on the mainland in a troop at Takasakiyama.[312]

The latest news from Japan is that Imo has by no means exhausted her powers, but has unleashed several additional cultural bombshells. Another of the foods provided at the stations is wheat, which the monkeys enjoy but find difficult to deal with once it has blown out of containers onto the sand. Imo was only three when she solved this dilemma by picking up mixed handfuls of sand and wheat and winnowing the grain by casting both into the sea. There the sand soon sank, leaving the wheat floating free on the surface, where it could easily be scooped up and eaten.[548] At the moment this subculture has spread only to Imo's immediate associates, but it will be fascinating to see what happens next. I personally wouldn't be surprised if, in her later years, Imo reinvented agriculture.

The relevance of this anecdote is that it suggests there may be mechanisms in evolution other than those governed by ordinary natural selection. I feel that there is such a thing as the Hundredth Monkey Phenomenon and that it might account for the way in which many memes, ideas, and fashions spread through our culture. It may be that when enough of us hold something to be true, it becomes true for everyone. Lawrence Blair says, "When a myth is shared by large numbers of people, it becomes a reality."[59] I'll happily add my one to the number sharing that notion, because it may be the only way we can ever hope to reach some sort of meaningful human consensus about the future, in the short time that now seems to be at our disposal.

I feel certain too that a contingent system exists and expresses itself in just this way. I suggest that the critical threshold for contingents was reached when our configuration, our form, brought them into the right kind of relationship in the right numbers for their own propagation. And I offer the notion that it was precisely their growing influence in evolution that created the tensions necessary for the brain to specialize in the way that has given us all the double-edged benefits of being conscious beings.

If there is anything in this notion, then at some point in our early evolution we will have gone through a Hundredth Monkey experience and, with any luck, it may have left its mark on the fossil record. We did, and it has.

There are very few traces of the first men, and behavior anyway leaves no fossil imprints, so we are forced to draw cultural and social inferences from the stones they left behind. Right through the entire period of the Lower Paleolithic, man made the same kinds of tools. No matter where you look today, on all sites of this age there are nothing but simple hand axes and basic chopping tools, virtually unchanged for more than a million years. Then suddenly, about a hundred thousand years ago, with the onset of the Mousterian culture, there is an explosive growth in style and complexity. And to me there seems little doubt that, in this first great quantum leap in the field of artifact design, we have evidence of the entry of a new factor into evolution.

I believe it was in this period that human language passed from a simple and probably highly ritualized system of sound to a true language full of complex conceptual invention—that this was the moment in time when we pulled several of the old threads together and began to build a full oral tradition. There is some evidence to suggest too that it was also at this time that man progressed from simple scravenging to become a truly efficient hunter. In order to bring down large animals without the aid of projectile weapons, early man must have developed persistence techniques, keeping his quarry constantly on the move, often for days on end, until it fell from exhaustion and could be butchered. To do this he needed not only a superior cooling system (it is possible that our nakedness and sweatiness derive directly from this time), but also a brain designed to hold the task constantly in mind, to anticipate results well into the future and to cope with a communication system loaded with enough symbols to keep a hunting group of separate individuals in close communal cooperation.[326]

Everything points to a comparatively short period of evolutionary ferment and rapid growth and change. The most interesting feature of the changes is that they were widespread and made on all locations at almost exactly the same time. This synchronicity is as difficult to explain as the sudden outbreak of pandemics in widely separated places, long before the advent of movement and trade. Unless one assumes the existence of a common factor, perhaps even the same factor—a new set of instructions.

I have already suggested that these instructions were new for us, but not necessarily new to us. They may have been dormant, waiting for the right conditions like the seeds of desert flowers through the long dry years between rains. And I think that we have evidence in the fossil record of just such a break in the weather of the mind.

The gross structure of an ape's brain is very much like our own, but there are radical differences in size, and in the proportion of body weight to brain weight. There is, however, no clear correlation between brain size and intelligence among modern man. If the brain is not deformed, and the volume is within the usual range from one to two thousand cubic centimeters, the owner can be expected to show normal human mental abilities. An individual whose cranium holds only twelve hundred cubic centimeters is not necessarily two-thirds as bright as one with eighteen hundred, but might just as probably be more intelligent. The modern average is fourteen hundred, but Anatole France made do, and won a Nobel Prize, with a mere thousand.

Herein lies a vital paradox. On the one hand, brain size does not affect human intelligence; while on the other hand, brain size is the very source of human intelligence. The solution to this apparent contradiction is very simple and very far-reaching.[314]

It seems that in the evolution of the brain, a critical size was reached where a small quantitative increase in brain substance resulted in a dramatic qualitative change in function. And it appears that this human Rubicon, the equivalent of the Hundredth Monkey in numbers of brain cells, stands at about seven hundred cubic centimeters.

Our body bulk lies between that of the chimp and the gorilla, and our physical activities are essentially similar to theirs. They function with brain volumes between four and six hundred cubic centimeters. So something like five hundred centimeters represents the required amount of brain necessary for maintaining the normal biological activities of the ape or human body. Human brains of this size are found only in microcephalic idiots, who may be physically quite normal, but mentally deficient in the sense that they have little or no ability to use or understand symbolic language. Such talents appear only with the addition of approximately another two hundred cubic centimeters of brain matter in excess of normal bodily needs.

We and our unique culture are products of this excess. Luxury products.

The Australopithecine apemen barely managed to reach five hundred cubic centimeters, but the recently discovered skull of "1470," a specimen of *Homo habilis* from Lake Rudolf in Kenya, crosses the divide with seven hundred and fifty cubic centimeters.[261] Why then did the cultural explosion wait for another three million years beyond his time? And why, if it be granted that virtual humanity is reached with a normal brain of this size, do modern brains average twice that volume? Since there is no marked intellectual improvement, what advantage is gained by this great increase? Nature is very careful and seldom does anything, let alone bringing about a major change of this kind, without good reason.

The answer lies, as it did with Japanese monkeys, in the inventiveness of children.

In the life of any modern man, the brain crosses the seven-hundred-cubic-centimeter threshold near the end of the first year after birth, and the child begins to use symbolic speech sometime during the following year, usually at about eighteen months, though it may understand several months earlier. The minimum brain mass therefore seems to be as vital an individual ontogeny as in the phylogeny of man.

Though *Homo habilis* as a species crossed the Rubicon and became a man, he could do so only late in life. His brain did not reach the critical mass until he was almost in his teens. Not until the age of about ten did his brain approach the volume and complexity of a modern one-year-old child's. By the time he was sexually mature, he had enjoyed the ability to use and understand symbols for only three or four years, whereas modern man reaches reproductive maturity with a minimum of twelve years of cultural experience. Given the comparatively low life expectancy of fossil man, this nine-year difference in the period of enculturation represents an appreciable portion of the life of most individuals. And such a short period of full cultural participation must have placed a severe limit on the kind of culture that could be transmitted from one generation to the next.[325]

So, despite the fact that all the necessary equipment was there perhaps as much as fifteen million years ago, contingent factors had to wait until *Homo habilis* developed before they could make themselves felt in an adult individual. And then there was another three-million-year pause before the sons of man grew up fast enough to give the new changes full cultural expression in the Mousterian explosion of activity.

When the genes discovered that there was fitness and survival value in having bigger brains, and that the greatest cultural advantages were conferred on those that gave their children the biggest brains first, the pressures were brought to bear. Brains grew, but heads grew too and eventually a point was reached when the skull became uncomfortably large for birth by a vertical animal with a poor pelvis. And so the system was redesigned to bring forward the moment of birth, shortening the pregnant period and producing an infant with a head of bearable size, but an almost unbearably long period of dependence. Though this was, and still is, a special and sometimes irksome human chore, it has, perhaps more than any other single factor, produced in us a strong sense of family and kinship.

We also contrived to slow down the rate of brain and head growth in the last months of pregnancy, and to make up for this deficit by an unparalleled burst of enlargement after birth. In the first year of a human baby's life, its brain trebles in size. With so much effort and energy concentrated up ahead, the rest of the body inevitably suffered and development lagged behind. Childhood was prolonged and enriched, and some infantile characteristics were retained right into maturity. Our teeth still erupt much later than those of our relatives. Our brows are smooth and juvenile, instead of ridged and armored, and some of our cranial sutures never close completely. We are, in effect, and in comparison to our ancestors and relatives, still children, with childish bodies and childlike faces. But we are driven by gigantic brains.

This "infantilization" or "fetalization" of man has long been recognized as an important factor in human evolution. The unkindest cut of all came from a French anthropologist who said that man "can be considered a gorilla foetus whose development and growth have been greatly retarded."[236] Now known, a little more kindly, as neoteny, the phenomenon helps to explain a number of our physical anomalies. Human ovaries, for instance, reach their full size at the age of about five, which is the time of sexual maturity in the living apes and presumably also in our extinct ancestors.[126] The rest of the human body, however, isn't ready for reproduction until many years later, being retarded by the action of hormones which play an important part in regulating the speed of our development. The anatomical facts are well established, but their significance doesn't yet seem to have been fully realized.

Ashley Montagu of Princeton rightly stresses the importance of the long learning period which a human child must undergo as a result of being born when, in effect, its gestation is only half complete.[382] David Jones and Doris Klein draw attention to the fact that man is the only species in the animal kingdom that continues to consume its infant food, milk, in its adult years. And they offer the view that "the preoccupation of man with intellectual pursuits constitutes a prolongation of the childlike attributes of learning, experimentation, and discovery."[289] They also point out that human sexual activity has taken on more of the playfulness of a baby animal, with the gratification of the individual assuming greater importance than the procreative end of the act. "Modern man has shown a tendency to convert everything he uses and does into toys and play; in doing so he has included sex and thus removed it progressively away from the area of instinct. In his increasing infantile quest for pleasure he has forced the removal of the shackles of many sexual taboos." The very use of the word "foreplay" suggests that this interpretation is probably the right one. But even these studies fail to draw what, for me at least, seems to be the most important psychological conclusions.

Human sexuality is precocious because it develops at about the same time as in our higher primate relatives, but in our species full development of the body in other ways is delayed. We go through a long latent period in which normal sexual impulses have to be repressed and controlled, until the individual is mature enough in other ways to take part in actual sexual activity.[459] This disharmony involving one of the strongest forces in nature has played a major part in the creation of the mind of man. I am temperamentally most unwilling to concede that there are any major qualitative differences between man and the rest of the animal world. But if there is one way in which we are different, I believe it is in this respect—that we have had to cope with a temporal imbalance in our physiology which has led to the development of some unique psychological mechanisms.

This is of course the basis of the revolution started by Sigmund Freud, and the reason for his fidelity to the sexual theory.

In the last years of the nineteenth century, Freud, then almost forty years old, was in private practice in Vienna as a consultant in nervous diseases. He had already experimented unsuccessfully with electrotherapy and hypnosis and was looking for a new approach, when he was asked to treat a woman suffering from hysterical paralysis. Hysteria is one of those strange fashions like the Victorian "vapors," which were once very common but seem now to be practically extinct. In the woman's case, only one hand was involved and the paralysis took the exact shape of a glove. Freud considered this "anatomical nonsense," since the muscles affected don't stop suddenly at the wrist.[193] His brilliant insight, which seems simple now only in retrospect, was that the paralysis must therefore be due to mental factors, which were not under the patient's conscious control. When he allowed her to become aware of this, the immediate result was that she regained the use of her hand; but the long-term effect of that first successful psychoanalytic experiment has been to alter our whole way of thinking about the mind.

"Hysterical patients," concluded Freud, "suffer mainly from reminiscences." Knowing that, he went on to wonder what it was that concealed some memories from consciousness, only to have them find expression in other ways. He turned for information to the most common experience of something beyond conscious control, to dreams. And he realized immediately that everybody had the same trouble with dreams, a tendency to forget them soon after waking. To Freud this suggested the existence of some sort of censor in the mind, which clamped down on dream information in the waking state, but was itself relaxed during sleep. But, in another stroke of genius, he detected the action of the censor even in dreams, condensing, changing, and distorting the subject matter so that every dream has a manifest content, which we can remember if we try; and a latent content, which has to be discovered by interpretation.[188]

Freud described the thought processes of the dream, in which time and space are distorted, and images and their symbols tend to become confused, as primary. He believed they represented a primitive instinctive precursor of the secondary processes, which involve logic and help us to deal with waking reality. And on the basis of this division he constructed the first working model of the mind, in which he distinguished between the conscious regions and another area, where there was little organization, contradictions abounded, and objects were frequently replaced by whole chains of association that had no rational basis. He called this second area the unconscious.

With these basic discoveries, Freud provided the first instrument for the scientific examination of the mind. With the technique of psychoanalysis (which very simply involves listening to what the subjects themselves have to say), the tool of dream interpretation, and the concept of the unconscious, he and a rapidly growing group of disciples began to explore an entirely new frontier. They worked out the rules as they went along, and it very soon became apparent that most of the conflicts which occurred—remember, they were using their equipment mainly on neurotic patients—were the result of tension caused by two basic forces which seemed to be seeking independent expression. These can be summarized in Eric Berne's observation that "man has a tendency to try to take what he wants when he wants it, and to destroy anything that gets in his way."[52] And what he mostly wants, and finds it hardest to get, is sex.

When Freud and his early co-workers started opening mental cupboards, it seemed that the first skeletons to come tumbling out all had genital problems. Right from the very beginning, psychoanalysis upset traditional concepts by asserting not only that adult sexual behavior has infantile precursors, but that the sex drives of children play a major role in determining adult personality. Findings in more recent years have tended to tone down this emphasis on the sexual theory, but it remains one of the most important formative factors.

One of Freud's first constructs was based on the sexual theory and called the Oedipus Complex, after the mythical king of Thebes who, unwittingly, killed his father and married his mother. According to classical theory, the complex consists of largely unconscious ideas and feelings revolving around the wish to possess the parent of the opposite sex and eliminate that of the same sex; and it emerges between the third and fifth year and is a universal phenomenon, responsible for much guilt. It remained a cornerstone of psychoanalytic theory until about 1930 and has been replaced since then by more mother-oriented ideas, but the fact remains that most children have two parents and grow up with some awareness of their sexual life together.

So we return to the knowledge that human beings are sexually precocious, prevented by custom and their own physical limitations from expressing these needs in the normal adult way. I believe that not nearly enough emphasis has been placed on the fact that this was probably the first time any organism had to conceal something from itself and that this repression could have been responsible in the first place for the development of an accumulation of covert impulses and ideas in an area of the mind now known as the personal unconscious. And I suggest that this development, by its very nature, being based on the unique and rapid growth of our cranial capacity, is confined to man.

But there is another side to the unconscious which Freud largely missed, though his reaction to Jung and to the looming phenomenon he described as "a black tide" suggests that he was not unaware of it, or of its potency.

Both consciousness and the human personal unconscious originate in experience, in environmental factors which shape the way in which a mind will develop. But the mind, no matter how complex it may be, seems to be dependent on a brain; and the form of the brain is strictly inherited. Every organism inherits a form that determines how it will react to the environment, and even determines what type of environment and experience it will have. "The mind of man is prefigured by evolution. Thus the individual is linked with his past, not only with the past of his infancy but more importantly with the past of the species, and before that with the long stretch of organic evolution."[232]

This is the ground on which Freud and Jung's friendship foundered. This placing of the personality as a whole within the evolutionary process was a landmark in the history of psychology and Jung's most magnificent achievement. With it he broke free from a strictly environmental determinism of the mind, and showed that evolution and heredity shaped not only the body, but all its manifestations. Jung called these deep waters of the mind the collective unconscious.

Robert Ornstein of the University of California suggests that the fundamental duality of our consciousness may be rooted in the fact that we possess two cerebral lobes,[413] and that Freud's distinction between the rational conscious and unreasonable unconscious might be explained by the different nature of left and right hemispheres. This is just possible if one considers only those parts of the unconscious material that are definitely of personal origin, such as memories or repressed individual experiences. But instinct can't be localized in that way on one side only of the brain.

The programs which determine how a fish swims or a locust flies, how a bird sings or a baby smiles, are all genetically composed and in vertebrates written twice, for safety's sake, once on each side of the brain. They are part of a long-term fixed species or racial memory and are open to negotiation only in the sense that individual personality may be altered by how much or how little these instructions get read. And this is where I believe that the contingent system comes in.

In addition to the conflict between consciousness and the personal unconscious which Freud considered in such detail, there is a separate tension between the personality and the collective unconscious. In Jung's words, "our personal psychology is just a thin skin, a ripple upon the ocean of Collective psychology. The powerful factor, the factor which changes our whole life, which changes the surface of our known world, which makes history, is collective psychology, and the collective unconscious moves according to laws entirely different from those of our consciousness."[302]

The contingent system is Jung's collective unconscious extended to include biological effects and factors, and given the possibility of a physical location connected in some way with the symbiotic nature of all nucleated cells. It interacts with consciousness and with the dictates of hereditary factors to produce the oceanic effects which I have identified as those of the Lifetide.

It is in this nonphysical, nonspatial, timeless no-man's-land of the psyche that ideas come into their own, that memes come to be and to influence being. And I suggest that among the archetypal memories and myths of our human origins are others that go much further back—back all the way to the ancestral clay, and perhaps even on into interstellar spaces. They have always been there, biding their time, waiting for the tide, surfacing now and then, sending shivers of mingled delight and dismay through phenotypes sentient enough to be sensitive to them, riding the images in our memories back down through the personal unconscious, fueling the conflicts with their own kind of feedback, floating up again to the surface in protean symbols, stirring things up, fomenting change.

This is the Lifetide—and it is time we came to terms with it.

# PART THREE

# The Tide in Our Affairs

It is important to have a secret, a premonition of
things unknown. It fills life with something im-
personal, a *numinosum*. A man who has never
experienced that has missed something important.
He must sense that he lives in a world which in
some respects is mysterious; that things happen
and can be experienced which remain inexplicable;
that not everything which happens can be anticipated.

The unexpected and the incredible belong in
this world. Only then is life whole.

—Carl Gustav Jung,
*Memories, Dreams, Reflections*

Julian Huxley said, "The first point to make about Darwin's theory is that it is no longer a theory, but a fact."[273]

Evolution is probably the most important scientific concept ever formulated. A measure of its greatness is that it underlies and makes sense of all biology, bringing order and connectedness to an enormous, almost embarrassing, abundance. It overflows too into sociology, anthropology, and philosophy with equivalent potency. But evolution must still be classified as a theory, because it is concerned with a phenomenon that never has been, and probably never will be, observed.

Nobody has ever seen evolution taking place. Not once in the history of biology has anyone been able to watch the simple differentiation of one species from another. So the most basic assumption of evolution, the existence of specific change itself, remains a theoretical construct. All that we have been able to do is to observe mutation within a species, and to assume that this is the mechanism which could account for the development of living organisms from specimens in the fossil record which look sufficiently like them to have been possible ancestors.

Evolution remains therefore what Jacques Monod calls "a second-order theory."[381] It is incapable of proof simply because we can't see it in its entirety. It takes too long, and our lives are so short.

The only way we can get any idea of a species, or of an individual member of a species, is to take it for granted that change is a fact of life, and to consider each organism, in Benjamin Burma's words, as "a succession of conformations of matter in time."[86] We need to try to think in four dimensions.

For instance, imagine an individual animal, something simple like a protozoan, which we'll assume for the sake of convenience answers to the name of Fred. Fred is a sexual animal and we can trace his origins back to the moment when his two parental gametes fused. At that instant he consisted of one cell, which we can refer to as Fred One. The next time we look for Fred One, we can't find him. He isn't there, but in his place is a stranger with two cells. Oh ho, we say, in our biological wisdom, Fred has undergone cell division, and we assume with confidence that, had we only been watching all the time, we would have been able to observe his continuous transformation into this two-celled stranger, who is obviously not Fred One. But since we are reasonably certain that he has some sort of close relationship to Fred One, we decide to call this new and more complex animal Fred Two.

As time passes, we discover that Fred Two is displaced in a similar way by Fred Three and he, in turn, by Fred Four, and so on—and in the end we are forced to admit that what we are seeing is not a rapid succession of different individuals taking turns to stand guard at the other end of our microscope tube, but a series of linked changes in one individual who is in the process of growing up. But what exactly is Fred? Obviously he is not the same thing at any two successive times. One moment he is Fred Seven Eleven, the next instant he is Fred Seven Twelve, and each manifestation of Fred is irrevocably different from any other that has ever existed before, or will ever come to exist in future. We can only make sense of Fredness by considering the whole ontogeny as a complex four-dimensional organism called Fred . . . One-plus-Two-plus-Three-plus-Four, etc.

If Fred as an adult has a simple spherical shape and all his adolescent forms are similar but smaller, we can look back in time through Fred's history and construct a four-dimensional perspective of him which will look something like an ice-cream cone tailing away at the sharp end to the single cell of Fred One. The same reasoning, of course, applies to Fred's parents and to all his other ancestors, so we could construct a four-dimensional view of the whole of Fred's species. And since there is a theoretical continuity of germ plasm from Fred and ourselves all the way back to the very first cells, we could, given the time and the necessary background, create a four-dimensional representation of all life that has ever existed on earth.

The historical shape of each living thing would be a sort of all-time phenotype for the entire species. It would be almost unimaginably complex and differ wildly from species to species. We don't have enough fossil information to make the roughest approximation of such a model for even one species, but it is possible to imagine what the four-dimensional shape of the entire biosphere would look like. It would be a fragile hollow sphere with a complex reticular skin just one-thousandth of its own diameter thick.

It might look a little like one of those magical photographs of the whole earth wreathed in wisps of cloud. But this wouldn't tell the complete story. It would be like looking at a single still picture, when what we need to do in order to see all life on earth at its own level is to view a movie film, taken perhaps from another star, of all its three billion years of development here.

The odd thing about movies is that the twenty-four single pictures which flow through the gate of the projector every second fuse into one continuous sequence. We see it that way because the human eye cannot form separate visual impressions of a number of scenes that replace each other this fast. Part of the limitation is in the brain and not the eye, because we know from experiments that viewers can quite unconsciously pick up information presented at twice that speed in just a fraction of one frame, despite the fact that they consciously experience this subliminal signal as no more than a flicker on the screen.

Even the human eye, however, is limited to events that take place no faster than fifty times a second. This factor, which is known as its critical fusion frequency or CFF, is what determines how we see the world and how we react to what we consider to be objective reality. It is ultimately this which determines the level at which we are conscious.

To take a simple example, a sheet of writing paper has many possible realities. To me it is a featureless rectangle until I begin to scribble on its blank face. But a faster organism with a CFF of less than a thousandth of a second, though it would see my hand crawling painfully slowly, might even be able to pick up the movement of molecules in the paper itself. While a slower being, with a CFF of an hour, wouldn't be aware of my hand at all, but would each hour on a good day see yet another page instantly and magically filled with closely reasoned manuscript.

Imagine that instead of a camera filming us from a nearby star, we are being observed by an awesome organism with a CFF equivalent to one earth week. Assuming that it, like us, has divided time into convenient portions about fifty CFFs long, one stellar second will roughly correspond to each earth year. And if it should in such a second happen to glance at us through some unimaginable telescope, this is what it would see: Earth's surface would be a blur, because it would be spinning at over three hundred revolutions per second. It would not be spherical, but would look instead like a diffuse cylinder stretched into an ellipse around our sun. And although this sounds paradoxical—you can prove it to your own satisfaction by constructing a simple model—the surface of the earth, the area on which we live, is on the inside of this cylinder.[517]

In 1818 a retired captain of the Ohio Infantry revived an old mystic notion that the surface of earth is concave, not convex, and that we live on the inside of our globe's surface with the sun at the center of this hollow sphere. By the turn of the century the "hollow earth hypothesis" had gathered quite a following, which seems to have included even the explorer Admiral Byrd, who flew over the North Pole in 1926 hoping to find there an escape route to the "other world." The idea was also taken seriously by some of the Nazi administration, who set up special radar apparatus on a Baltic island to test the theory.[426]

All the geophysical evidence denies this possibility. The earth is obviously a solid sphere with a possibly molten core, but clearly not hollow. Not at our level of reality anyway. But alter the perspective slightly by changing nothing more than the viewpoint and the sensitivity of the organs used to collect information, and the entire model changes. We are dealing immediately with another reality and a separate consciousness of the same thing.

I suggest that the contingent system operates on such a long-term four-dimensional basis. If it has any awareness at all, if it expresses any interests in us, they will be of this kind. And I suspect that in its action on, and interaction with, our system, we will find possible solutions to some of our mysteries, and premonitions of many more things still unknown.

# CHAPTER SEVEN

# COLLUSION: Being Conscious

Evolution makes mistakes. As Arthur Koestler puts it, "For every existing species hundreds have perished in the past; the fossil record is a waste-basket of the Chief Designer's discarded models."[322] True, but evolution also gets things incredibly, almst unbelievably, right.

The transparent cornea of our eye could hardly have evolved through progressive trial and error by natural selection. You can either see through it, or you can't. Such an innovation has to be right the first time, or else it just doesn't happen again, because the blind owner gets eaten. Darwin himself admitted that the perfection of the vertebrate eye sent cold shivers down his spine.

Something else which disturbed him was the elaborate life cycle of certain insects, "in which we cannot see how an instinct could possibly have originated" and "in which no intermediate gradations are known to exist."[122] The lyric French entomologist Jean Henri Fabre, who worked all his life alone on the sandy stretches of southern France, put his finger right on such a sore spot in the theory of natural selection, when he raised the controversial matter of the giant wasps.[170]

Adult wasps are vegetarian, but the larvae of many are carnivores. So in these species the survival of the young depends on the mother's correct choice of food which she herself does not eat. This much can be under instinctive control, but there are refinements in the relationship between the predatory wasps and their prey which are almost impossible to squeeze into an evolutionary or instinctive model.

To take just one example, the wasp *Pepsis marginata* feeds its young only on the tarantula *Cyrtopholis portoricae*.[431] The female wasp produces very few eggs, but for each one she has to provide an adult tarantula, alive but paralyzed. When an egg in her ovary is almost ready to be laid, she goes out hunting, flying low over the ground late on a sunny afternoon looking for a spider out early in search of its own insect food. The tarantula has poor sight and little or no sense of hearing, relying on an extremely delicate sense of touch to locate its prey. The lightest contact with any of the body hair on a hungry tarantula and the spider whirls and sinks its long fangs into any cricket or millipede that gets too close. Yet when spider and wasp meet, and the wasp starts to explore with her antennae to make certain that she is dealing with the right species, the tarantula does nothing. The wasp crawls under the spider and even walks all over it without evoking any hostile response. If the molestation is too great and too persistent, the tarantula sometimes rises up on all eight legs as though it were standing on stilts, but otherwise calmly awaits its fate. "All is arranged," muses Loren Eiseley, "in such a manner as to suggest the victim possesses an innate awareness of his role, but cannot evade it."[155]

Meanwhile the wasp moves off a few inches to dig its waiting victim's grave. Working vigorously with legs and mouthparts, she digs a hole about ten inches deep and slightly wider than the spider, popping her head out of the excavation every now and then to make sure that the tarantula is still there. Usually, and unaccountably, it is, and when the grave is ready, the wasp returns to complete her ghastly enterprise. First she feels the spider all over once again with her antennae and then slides underneath on her back, working with her wings to get into the right position for a shot at the vital spot. She can penetrate the spider's horny exoskeleton only at the soft hingeing membrane where the legs join the body, and only if she stings with surgical precision to the right depth, at the right angle, and in precisely the right place can she be certain of locating the one nerve center which will stun the spider without killing it. And during all this maneuvering, which can last for several minutes, the tarantula makes no move to save itself.

Finally the wasp jabs and the spider tries a desperate but vain defense. The two roll over and over on the ground, but the outcome is always the same. The tarantula falls paralyzed on its back. The wasp drags it by one leg down into the waiting tomb, where she does another remarkable thing. She packs her big hairy larder so masterfully into the hole that even if it were by some chance to recover, it could never dig its own way out. Each one of the eight huge limbs is literally handcuffed to the earth. Then she lays one egg, attaches it to the side of the spider's abdomen with a sticky secretion, fills in the grave, and leaves.

But the extraordinary story doesn't end there. When the wasp larva hatches, it is many times smaller than its helpless victim, and totally dependent on it. During the long weeks of development it will have no other food and no water, and so, working to a complex and gruesome culinary program, it proceeds to consume the tarantula piece by piece, keeping it alive and fresh by saving the vital organs until last. By the time it has completed its gargantuan meal, and is ready to burst out of the tomb carrying its own surgical instrument, and a map of operations to be performed on another tarantula, nothing remains of the first one but its indigestible chitinous skeleton.

Mutations in structure or behavior are, argues Jacques Monod, "drawn from the realms of pure chance."[380] But if that were so, we would expect that the tarantula would also by now have accidentally happened on a defense against its predator. Instead we have a situation in which a spider, quite capable of defending itself against or even killing a wasp, allows the insect to paralyze it. And we have a wasp that has an uncanny knowledge of the exact location of the nerve centers in its prey. Planted anywhere else, the sting will either kill the spider, rendering it useless as a food store, or fail to have any effect except the probable death of the wasp by retaliation. In neither case is there scope for natural selection. There are no degrees of success in this endeavor. It is an all-or-nothing situation. You can't go out practicing hypodermic skills on poisonous spiders twice your size. You must get it right the first time.

Evolution asserts that spectacular adaptations have their origins in myriads of minute mutations, the vast majority of which are harmful to the organism; and that natural selection acts like a ratchet, preserving each useful mutation while new changes are being tried out. This implies that the wasp's master chart of surgery was not always perfect. But a surgeon cannot learn his trade by indiscriminately chasing and slashing at potential patients with a scalpel. This evolutionary marvel in the wasp could not come about by the sort of slow selection that we know from the fossil record worked on the horse's ancestors to give the living members their larger size and greater speed. In the wasp the entire pattern has to work immediately, or the species becomes extinct. And how could a pattern as complex as this come about in isolation, on the off chance, without actually being used? Because, before it was complete in all its details, it could not be used at all.

Examples of this kind have bothered biologists ever since the Darwinian theory was first published. Fabre said, "It is not in chance that we will find the key to such harmonies," and he chose instead to accept that some mysteries were incapable of solution. "Man grappling with reality," he concluded, "fails to find a serious explanation of anything whatsoever that he sees."[170] Loren Eiseley, late in his life, came to a similar conclusion. "In the world there is nothing to explain the world. Nothing to explain the necessity of life, nothing to explain the hunger of the elements to become life, nothing to explain why the stolid realm of rock and soil and mineral should diversify itself into beauty, terror and uncertainty. To bring organic novelty into existence, to create pain, injustice, joy demands more than we can discern in the nature that we analyze so completely . . . I am simply baffled. I know these creatures have been shaped in the cellars of time. It is the method that troubles me."[155]

And me. One ends up always falling back on the idea of some sort of design in nature, which implies the existence of a Designer. That may be the final and perfectly reasonable solution, but it is a little embarrassing for a scientist because it is a theory incapable of refutation. As Karl Popper puts it, "falsifiability, or refutability, is a criterion of the scientific status of a theory."[436] You have to be able to test a theory. You have to be able to prove that it is right or wrong. An explanation which explains everything, explains nothing. An explain-all is no more credible than a cure-all. Both are bad science and lousy logic.

Which is why I become increasingly fond of the theory of a contingent system. It goes some way toward establishing numinous forces at a more physical, substantial level where they, or at least their direct effects, can be examined and tested. It provides a handle on one of the biggest problems posed by both wasps and man, which is to find some way of accounting for the apparent inventiveness of evolution. We need to know how it can cross species lines and coordinate changes in the biosphere as a whole. And the major difficulty in the past, which has led even men of Eiseley's stature to give up all attempts to find an explanation, has been a block in our thinking about the rate of mutation.

Traditionally a species is thought of as being more or less "at rest" most of the time. It is presumed to be well adapted to its current environment and, if natural selection acts on it at all, it is to maintain the species characteristics rather than to change them. Occasionally, of course, there might be some change in the environment, such as an ice age or the appearance of a new predator. And when this happens, it imposes new selective pressures, and the species responds to this by a sudden burst of evolution, or else it becomes extinct. Now that we know how little of the genetic material is directly involved in active service, it becomes easier to understand how sudden and appropriate changes can be made. The organism doesn't have to wait for a new mutation to arise to meet the new challenge, but simply falls back on the large amount of latent variability it has been accumulating during its peaceful phase.

If you take a population of laboratory mice and allow only the largest ones to breed, you can by this sort of artificial selection speed up the natural increase in average size by as much as a hundred thousand times. This is precisely what man has done in order to produce, in a very short time, the wide variety of breeds now found among domestic dogs. But John Maynard Smith points out that this is not speeding up the natural rate of evolution, because the changes slow down and stop when the initial supply of genetic variability has been used up.[493] It is very rare now for an entirely new breed of dog to appear on the market.

The block has been our inability to appreciate that no species, no matter how well adapted it might be, can afford to relax for a moment. There is no such thing as a stable ecology. No living thing can be at rest, because the main feature in the environment of all species is other species, which constantly try to eat, avoid, compete, or combine with it. And when any species in any ecosystem makes an evolutionary advance, this is experienced by one or more of the others as a deterioration of their environment. Their prey becomes more difficult to find, or their predators more difficult to escape. So these species evolve in their turn and cause comparable environmental deterioration for still other species. And so the ripples in the pool expand until everyone is affected in some way by every change. The net result is that every species alive anywhere in the world is probably evolving as fast as it can, simply to keep up with all the others.[560] Our world, it seems, is governed largely by what some biologists are beginning to describe as the Red Queen Hypothesis. It was the Red Queen, you will remember, who told Alice, "Here, you see, it takes all the running you can do to keep in the same place." Which makes it all sound very hectic. It is, but there are advantages as well as disadvantages in being interconnected in this way.

All cuckoos have their favorite hosts. In Europe the great spotted cuckoo *Clamator glandarius* picks mainly on the magpie *Pica pica*, returning every spring to sneak one of its eggs into each of half a dozen unwary magpies' nests. Sometimes a magpie spots the intrusion and tosses the offending egg out, but over millions of years of experimental freeloading, the cuckoos have developed eggs which are remarkably successful mimics of the host's own productions. In 1973 three young biologists began an experimental study of brood parasitism in the Donana Reserve in southwestern Spain.[6] They introduced a variety of egg models and eggs from several other species into hundreds of magpie nests, to find out precisely which stimuli the hosts most often responded to. And, toward the end of their program, they also tried introducing the featherless chicks of swallows, sparrows, jackdaws, and starlings. They found, as expected, that the magpies most strongly rejected eggs which looked least like their own, but that very few of the foreign chicks were turned out. This also is not surprising, since it is eggs and not chicks that the parasitic cuckoos deposit in their nests, so there has in the past been little need for the magpies to have developed an anti-chick device. In any case, the cuckoos have anticipated them in this too, by producing chicks that mimic and even exaggerate the normal nest behavior of the magpie's own brood. However, one of the adopted chicks forced on the magpies by the researchers did produce an absolutely mind-boggling evolutionary innovation. They report this incident as a brief aside in their paper, but it is a biological discovery in its own way as momentous as man's first use of fire.

The chick in question was one belonging to the European swallow *Hirundo rustica*, which is not normally parasitized by cuckoos. It was put into a magpie nest already occupied by three magpie eggs. The next day the researchers noticed that one of the eggs had fallen onto the ground below the nest. It was not damaged, so they picked it up, replaced it, and watched. "We were able to observe that the swallow chick repeatedly loaded an egg onto its back whilst climbing onto the edge of the nest, dropping the egg on the ground below the nest twice in front of our eyes."[6]

There is revealed, even in this bald statement, a little of the incredulity which all biologists must feel about this kind of event. The baby swallow was behaving exactly like a baby cuckoo, balancing the egg on its back between the wing stubs, and walking backward up the side of the nest until the egg toppled out. But how could it ever learn to do a thing like this? Baby swallows are not accustomed to finding themselves in magpies' nests, or any other nest except their own. Could the behavior be an anti-cuckoo adaptation? Something brewed up in the swallow's gene pool as a sort of preemptive strike against possible future cuckoo parasitism? Or perhaps a relic left over from a time when cuckoos did once try to move in on the swallow? According to this theory, the baby swallow could have been reacting to the magpie egg as though it belonged to a cuckoo, simply because it was larger. But if the swallow chick can tell the difference, then so can a swallow parent, and it would be far easier for an adult to remove the egg. The fact that the difficult and complex task of egg rejection was seen being performed by a normally weak and helpless baby swallow means that the cuckoo theory must be abandoned. Something else, something much more extraordinary, is going on.

I suggest that in this behavior we are seeing the contingent system in action. There is room in the normal excess DNA for even a swallow to carry unused cuckoo instructions, just in case. There may even be ostrich and great auk and dodo plans filed away somewhere in there; but no system of natural selection, not even one given unlimited scope and unlimited time to work in, could ever arrive at all possible conclusions purely by chance. There has to be some system of communication of instruction between species. Our new awareness of the free-floating packets of DNA in viruses, and of the existence of viral-host genetic exchange, provides a possible mechanism; but it is one which still lacks coordination and design. I believe the contingent system provides just that. It is common to all organisms with nucleate cells. It is almost as old as life itself. And it operates with a different sense of time and space, with its own peculiar form of subjectivity, its own intelligence. The contingent system is the collective unconsciousness of every living thing.

There seems to be almost no limit to the freedom the system can exercise in borrowing and exchanging patterns even between radically different organisms in totally distinct phyla.

In the Amazon, invention runs rife and there are some extraordinary adaptations, but none more wonderful than the planthopper *Laternaria servillei*. This is an insect three or four inches long, a relative of the cicadas and the aphids. It feeds on plant juices down near the water's edge and normally sits with its mottled brown wings folded back over the length of its body like a horny case. The head is enormously bulbous and elongated, drawn out into a snout with a nasal prominence in front and large, false eye bumps behind, each of which has a white mark in exactly the right place to simulate the glint of light reflected from a real vertebrate eye. Along the sides of the "snout" is a groove that makes it look like a partly opened mouth, and staggered along this are a series of white false teeth, rendered not only in color but in perfect bas-relief. The total effect is an incredibly lifelike imitation of an alligator, small, but perfect in every detail.[440]

I have discovered many of these planthoppers dozing on the banks of muddy tributaries and, despite the fact that I know them for what they are, and have myself little to fear anyway from a four-inch alligator, find that it takes a conscious effort of will to reach out and grasp one by the head. They embody an essence of alligatorness which is actually terrifying, and was probably intended to be. The usual predators of planthoppers are birds such as herons and kiskadees, which operate along the river margins, where they soon learn to watch out for lurking alligators. And this caution seems to extend to the alligator mimic, because birds are more concerned with color and shape and distinguish improbabilities in these parameters more readily then differences in size. An oystercatcher will, for instance, try to incubate a giant egg almost the size of a football, as long as it has the usual mottled markings.[542]

This astounding example of insect sculpture can hardly be the product of sheer chance. The mechanism of evolution in this planthopper must have been at work with some reference to the alligator. And it is impossible to imagine any way this could come about except by assuming some sort of informational flux in that part of the forest, an interchange of pattern and instruction which gives normal Darwinian natural selection something concrete to work on. I contend that situations like this demand the existence of a contingent system. Without it, or without something very like the system I have suggested, we are left floundering in the face of meaningful, appropriate, inventive evolutionary adaptation of the kind shown by *Laternaria*. With it, we have an instrument which provides at least an element of understanding of this, and possibly a great deal more.

I am well aware, in setting up this still rather mystical contingent system, of falling into an old trap. One doesn't come to a new understanding simply by giving an old problem a new name. A fellow biologist, John Randall, delineates the difficulty rather nicely in his remark, "Around the frontiers of advancing science there are always areas where our understanding is incomplete, and there are always some people who will seize upon these areas as a justification for introducing some kind of metaphysical entity—a god of the gaps."[445] In fairness to him, I must add that he is assuming this mechanistic stance as part of a debate and that he has, like me, somewhat vitalistic leanings. There have always been these two polar approaches to the problem of the nature of life. The mechanist believes that all life can be described in terms of physical and chemical equations, and that science will one day succeed in doing this. The vitalist believes that living matter possesses some extra magic ingredient, a vital force or essence that cannot easily be reduced to mechanical components.

Ever since Aristotle, who invented the term "entelechy" (which means "having an inner purpose") to describe the directedness of life, the pendulum has swung between mechanistic and vitalist extremes. The last flourish of vitalism took place around the turn of the century, spearheaded by the young German biologist Hans Driesch.[137] But it was routed by the triumph of organic chemistry and, more recently, by molecular biology. Properties which were once thought to be characteristic of life have been shown to be fully explicable in physical and chemical terms, and some have even been demonstrated with nonliving materials in test tubes. The mystery of reproduction has been analysed away by the elucidation of DNA, and many apparently purposive patterns have been duplicated in cybernetic systems. It seems that there is little left that requires the existence of any kind of "vital force," and science, even the science of life itself, has become almost totally mechanistic.

Vitalism, if not actually a scientific sin, is seen at best as a delusion. The geneticist Theodosius Dobzhansky refers to it as a "sham solution of biological riddles."[131] And Gordon Rattray Taylor describes it as a "question-begging explanation," which attempts to explore the mysterious "by postulating some ill-defined force capable, by definition, of doing whatever was seen to be done."[533] There is no doubt that mechanistic reduction is a very successful way of exploring some aspects of the world. Yet each time science attempts once and for all to exorcise the persistent ghost in the mechanism, it seems to pop up somewhere else. Arthur Koestler carries on with his guerrilla war against all attempts to reduce life to the level of conditioned reflexes;[320] and an increasing number of life scientists are joining him in expressing dissatisfaction with purely mechanical explanations. Ludwig von Bertalanffy of the University of New York notes that if we had not made the initial mistake of thinking of organisms as machines, the ghost probably wouldn't have appeared in the first place.[563]

The most productive approach seems to be an organismic one which looks not at the separate components, but at the living system as a whole, uniting mechanistic and vitalistic insights in an attempt to appreciate the complete picture. It is in this sense that I see the contingent system, which we may be able to explain through the particularities of organelles and cell communities, but which we will never understand until we can see the biosphere in its entirety. And this may involve stretching the imagination to allow an extraordinary four-dimensional perspective.

With some of the discoveries in quantum physics, it is gradually dawning on us that science has been so successful simply because it has abstracted from the world only those aspects of reality which are capable of being analyzed in the mechanistic way. We have in effect been reading only those books which support our point of view. The physicist David Bohm says, "Thinking within a fixed circle of ideas tends to restrict the questions to a limited field. And, if one's questions stay in a limited field, so also do the answers."[63]

So I persist in my belief that the most valuable glimpses we get of the true nature of life lie in the ill-defined area of the so-called supernatural. It is on rare occasions and in isolated incidents that we get the chance to look at ourselves with fresh eyes as part of what Albert Einstein called "an intelligence of such superiority that, compared with it, all the systematic thinking and acting of human beings is an utterly insignificant reflection."[154] We and the planthopper and the baby swallow all have direct access to this intelligence, because we are reflections in its pool, part of the tidal system.

Helmut Schmidt of Duke University has been involved in several pioneering attempts to track down elusive phenomena. Most of his experiments involve the use of sophisticated electronic apparatus with human subjects, but he has recently tried out one piece of equipment on a cat. Schmidt linked a binary random-number generator in his home to a heat lamp in a garden shed, so that the light turned on and off at strictly random intervals. When the shed was empty, it did just that, showing no tendency to generate unusual sequences and keeping the light on exactly half of the time. But when a cat was confined to the shed in cold weather, the machine kept the warm lamp in the unheated room on far longer than could be expected according to chance alone. Somehow the presence of the cat made a difference.[476]

In atomic physics it is no longer possible to talk about the properties of an object like the number generator as such. They are meaningful only in the context of that object's interaction with an observer, with consciousness. The crucial feature of this new understanding is that the observer is directly involved, even to the extent of influencing the properties of the object. Schmidt himself was obviously included in his own experiment, and it may have been he who influenced the machine; but if the results continued to be biased even when he didn't know whether or not the cat was present, we are forced to assume that the cat itself was in some way responsible for controlling its own environment.

The effects are even more pronounced when more than one organism is involved.

At the University of Utrecht, it is the mice that play. Sybo Schouten began by training ten mice to press a lever in whichever half of their cage an indicator light went on. If the mouse got it right, it received a drop of water as a reward. If it got it wrong, nothing happened. When all the mice were properly trained, Schouten put one in a cage containing lamps but no levers, and another in a cage several rooms away with levers but no lamps. Watery rewards appeared simultaneously in both cages if the lighting of the lamp in one, and the pressing of the lever in the other, coincided. The timing of the lamp switch was controlled by a binary random selector and the results of the experiment were recorded automatically on punched tape, so that no humans were directly involved.[477]

In the first series of experiments, several of the mouse pairs consistently produced scores greater than could be accounted for by chance alone. This seems to show that when the lamp lit in the cage of the first mouse, it was able somehow to transmit this information to the second thirsty mouse, who then pressed the appropriate lever to give them both the desired reward. In an attempt to distinguish between the possibility of this sort of communication and prediction, in which the second mouse was just guessing correctly when to press a lever, Schouten ran a second series of tests in which he simply left the first cage empty. With no mouse there to see the lamp and broadcast news of its illumination, a number of mice in the second cage still succeeded in producing better than chance scores. But the interesting thing is that they were not the same mice that scored well in the first series. There are, it seems, in mice and men, some subjects with telepathic talents, and others whose abilities seem to be more clairvoyant. Our connections with, or sensitivity to, the contingent system are by no means standardized. There is room in the interreaction for the exercise of will.

The last and greatest of the mechanistic strongholds has been provided by molecular biology, which seems to have succeeded in cracking a code so simple, and yet so potentially complex, that it can act like a giant computer, holding information and instruction comprehensive enough to account for everything that happens. The genetic system does have this potency and yet the latest discoveries seem to indicate that nothing in there is as linear and tidy as we would have liked. It is a dynamic system in a state of considerable flux and can be influenced by other factors. The capacity of DNA for storing and processing data is immense, but I suggest that on its own it cannot result in intelligence greater than that we could expect from any machine. True mind depends, I contend, on the creative intrusion of a second system, which sets up subtle interference patterns between the two. It is in these patterns, which are in effect something like holographic effects— apparently three-dimensional and yet totally unreal—that mind and will exist.

If this is so, then it ought to be possible to demonstrate that mind has control not only over inanimate matter in mechanical generators, and not just over physiological effects in a living body, but even over the genetic system itself.

One of the nastiest of all genetic defects known in man is a hereditary condition called congenital ichthyosiform erythrodermia, Brocq's disease. It is characterized by a malfunction of the sebaceous and sweat glands which results in an excessive growth of the epidermis, the outer layer of the skin, so that a dark, horny covering develops like the scales of a primitive fish. This sometimes sheds or sloughs in fragments in a way which suggests that the disease might even be a throwback to an earlier phylogenetic state. It is an appalling and totally disfiguring condition that normally continues throughout the patient's life, which is apt as a result to be short. And until 1951 it was considered to be incurable.

In that year a sixteen-year-old boy with advanced Brocq's disease was referred as a last resort to A. A. Mason of the Queen Victoria Hospital in London. The boy had a black horny layer covering his entire body except for his head. "To the touch the skin felt as hard as normal fingernail, and was so inelastic that any attempt at bending resulted in a crack in the surface, which would then ooze blood-stained serum."[369] These cracks subsequently became infected and the net result was a smell so objectionable that the boy couldn't even be sent to school. He had been born in this state and subject all his life to a variety of treatments which culminated in an attempt to graft healthy skin from his neck onto some of the affected areas, but though the grafts took, they quickly became affected in the same way.

Mason is a skilled hypnotherapist, and as soon as he discovered that the boy was a deep-trance hypnotic subject, he began treatment by suggestion. During their first session on February 10, 1951, they confined their attention to the left arm. On February 15, the horny layer stripped clear away, leaving unblemished skin which was pink and soft. "At the end of ten days the arm was completely clear from shoulder to wrist." Then they went on to deal with the other arm and the rest of the boy's body in the same way. After several months the symptoms of the disease had almost gone and he was able for the first time to lead a normal life. He became apprenticed to an electrician and "In a five year follow up, the patient was still alive and well and free from his dreadful inheritance."[58]

There is no question that Brocq's disease is congenital and under control of a gene whose dominance seems to depend partially on sex-linked factors. It is, in the words of a specialist, "as much an anatomical maldevelopment as is clubfoot."[55] And yet Mason, by invoking only the assistance of the boy's mind, was able to effect a cure. He succeeded, if not in altering the genetic instructions themselves, then at least in counter-manding them, which amounts to the same thing. DNA doesn't necessarily have the last word.

Since that time there have been a number of similar cases. Congenital linear naevus, a protracted kind of birthmark, has been totally removed by suggestion.[212] And a patient with congenital pachyonychia, an almost hooflike enlargement of the nails and feet, has been restored to normal dimensions.[388] There has been direct and external influence brought to bear on strictly genetic mechanisms. Mind stalks, it seems, even in the sacred hunting grounds of molecular biology.

The vital conclusion that can be drawn from this clinical evidence is that the therapy in each case involved contact with the patient's personal unconscious. The conscious mind on its own seems to be incapable of relieving even psychosomatic problems. Nobody ever cured asthma or eczema simply by telling the symptoms to go away. But unconscious action, most easily initiated under hypnosis, has relieved hypertension, peptic ulcers, colitis, hay fever, allergy, psoriasis, warts, shingles, and even tuberculosis.[57] Why is this? Well, if it is true, as I have already suggested, that awareness can be traced directly back to the first time a cell, or a collection of cells, was able to distinguish "me" from "not-me" and that it was this ability which grew into a fully functioning immune system responsible for maintaining the integrity of the body, then the personal unconscious could be the area in which that discriminating capacity still survives. It simply got shouldered aside by the new burgeoning consciousness and exists today as a sort of "body-mind" in cooperation with the newer, more independent "brain-mind." And being close to hand, it became also a ready and convenient receptacle for what Jung calls "the sediment of experience," which sifts safely down to areas where it can be sealed off like radioactive waste.[293]

This is a very simplistic portrayal of the respective roles of consciousness and the personal unconscious, but I believe it is accurate in that it creates even at this level the need for a third system, an intermediary responsible for the sympathy of all things—a comprehensive all-life enlargement of Aldous Huxley's "mind at large" and Jung's "collective unconscious." This is the consequence of the contention between genetic and contingent systems. This is the substance of the Lifetide.

Arthur Grimble, that most delightful and enlightened of all colonial administrators, tells of an occasion on Butaritari, the northernmost atoll of the Gilbert Islands, when he took part in an extraordinary interspecific exchange.[225] On a day arranged weeks in advance, he joined the people of Kuma village while their hereditary porpoise caller lay alone in a small grass hut on the beach, feet to the west, summoning the dolphin in a dream. In the late afternoon, "a strangled howl burst from the dreamer's hut. I jumped round to see his cumbrous body come hurtling head first through the torn screens . . . A roar went up from the village, 'They come, they come!' I found myself rushing helter-skelter with a thousand others into the shallows, bawling at the top of my voice that our friends from the west were coming." Fifty yards short of the reef the people stopped and watched as a shoal of porpoises cut through the surf outside. Then they surged through a gap and moved in toward the waiting crowd "in extended order with spaces of two or three yards between them, as far as my eye could reach. So slowly they came, they seemed to be hung in a trance. Their leader drifted in hard by the dreamer's legs. He turned without a word to walk beside it as it idled towards the shallows . . . A babble of quiet talk sprang up . . . The villages were welcoming their guests ashore with crooning words. Only men were walking beside them; the women and children followed in their wake, clapping their hands softly in the rhythm of a dance. As we approached the emerald shallows, the keels of the creatures began to take the sand; they flapped gently as if asking for help. The men leaned down to throw their arms around the great barrels and ease them over the ridges. They showed no least sign of alarm. It was as if their single wish was to get to the beach."[224]

Every now and then various species of whale and dolphin do manage to beach themselves, usually in groups, as though taking part in a suicide pact. Explanations have been offered for this bizarre behavior, ranging from freak storms to disorientation produced by parasites in the inner ear. But all these theories wear the uncomfortable air of city-suited guests at a fancy dress party. They just don't fit in with what we know of the facts and what we are beginning to learn of cetacean mentality. And they certainly can't account for a beaching like the one Grimble observed, by arrangement, and on a day he himself had chosen at random several weeks earlier.

On a remote Indonesian island in the Banda Sea, I once lived for a while with the people of a fishing village. The expert among them, a man they called the *djuru*, was justly famous for his ability to locate and identify fish underwater, even in the dark, just by putting his head under and listening. On one occasion he even detected, half an hour in advance of its arrival, an approaching tidal wave.[573] This much can be accomplished by sensitivity and training, but there was one incident which occurred during the time I spent with him that I can't file away in a mechanistic pigeonhole quite so easily.

In the month before the northwest monsoon, female marine turtles haul themselves laboriously up the beach on the seaward side of the island. They always come at night, usually in the dark of the moon, making their way up beyond the high-tide line and, after testing several places, digging a nest pit in soft sand among the trailing creepers of beach morning glory on the edge of the forest. I watched them many times, never tiring of the cycle and always moved by the effort involved, by the heavy breathing and the secretions that flowed like real tears down their scaly cheeks. Once they started to dig, nothing could distract them; the whole sequence of behavior, once set in train, had to run to its natural conclusion. But the turtles were easily frightened during the half-hour they spent near the water's edge, watching and listening to make sure the coast was clear. I had to lie very still up under the tree canopy, not making a sound nor letting my silhouette be seen until each huge body had hauled itself gasping up to the soft dry sand.

Most of the turtles were green *Chelonia mydas*, with an occasional hawksbill *Eretmochelys imbricata* coming up just before dawn. I enjoyed them all, but kept on going out night after night in the hopes of seeing the giant of all the sea turtles, the rare leatherback *Dermochelys coriacea*. These are known to come up in groups on the eastern coast of Malaysia at Trengganu, but they are sometimes solitary nesters, and seeing one was just possible. But I searched each night in vain, until I mentioned my interest to the *djuru*. He recognized *penju kulit* from my description and promised to show me one. No, more than that, his actual words were "I will dream one for you."

Almost a month later, he came early one morning to the school where I was teaching and said that all was arranged for later that day. From the way he spoke, I assumed that a leatherback must have been caught in the nets and was being kept for me to see. But I was wrong. In the middle of the afternoon we went together, not to the ocean beach, but down to a sheltered corner inside the reef. The tide was at its peak and he took me out to a place called Batu Jari, the "toe rock," which marked the limit of a lava flow from the island's main volcano. From there it was possible to look directly down into a deepwater channel where coral banks dropped away to the smooth floor of the great blue hall of the lagoon. I sat up on a raised shelf, as I had many times before, watching the kaleidoscope of colorful reef fish darting in and out of crevices in the coral, while the *djuru* crouched at the water's edge with both hands immersed, working the water with his fingers as though he were playing it like a piano. This went on for twenty minutes or more until my attention was distracted by the sight of a frigate bird swooping down at the surface of the lagoon about a hundred yards out, doing, I assumed, a little of his own fishing for a change instead of stealing from the hapless boobies. As he came closer, I could see what it was that interested him. A huge dark shape just beneath the surface was gliding toward us like the shadow of a cloud. I thought at first it was a manta ray, but the movement wasn't right. A shark, perhaps? Maybe a basking shark? A whale? Despite the reason for our being there, it wasn't until a shiny olive-green back with longitudinal flutes broke the surface behind a chunky bullet head that I realized it was *penju kulit* herself, a giant leatherback turtle, larger than I had dreamed it could ever be. The *djuru* had seen it too, but he carried on with his dabbling, starting now to match his movement with a quiet chant in the old tongue. The turtle seemed to see me, almost dancing in my excitement, when she was about twenty feet away, and she turned suddenly on her side so her belly flashed white, and winged away into the deep. But she didn't leave. Three, four, five times she swept up and down the channel in front of us, each time a little closer and a little more calmly. Finally she stopped directly opposite the *djuru* and came drifting slowly to the surface to breathe. When she breached, her head was only three feet from his hands. He kept his right hand underwater and stretched out with his left, palm down, toward her. And while I watched, incredulously, she lifted her beak deliberately up toward him, and twice, as gently as a suckling calf, nibbled at his fingers. Then she turned and swam with swift clean strokes of her flippers straight out toward the gates of the lagoon and into the open sea.

Now, at this distance, the whole incident has a dreamlike quality. I believe it happened just the way I have described it, but it must remain like Grimble's account, an anecdote without scientific validity. And yet if one searches the literature, or better still talks to curious naturalists who have had comparable experiences but hesitate to commit them to paper, something like a series begins to build up. Science, even natural science, in its determination to be objective, and in its insistence on documentary evidence and repeatability, may be missing something.

In 1940 the twelve-year-old son of a county sheriff in West Virginia was taken a hundred and twenty miles to the Myers Memorial Hospital at Philippi for an operation.[441] One dark, snowy night, about a week after his arrival, he heard a fluttering at the window of his hospital room. He called a nurse and told her there was a bird trying to get in. To humor the boy, she opened the window and a pigeon came right in. He immediately recognized it as his personal pet. He told the nurse to look for a ring on its leg carrying the number 167. She did, and there was. He was allowed to keep it in a box near his bed, and when his parents came to visit a few days later, they confirmed that it was indeed his bird and had not been seen around the house for several days after he was admitted to the hospital. So it hadn't been brought with him, or simply followed the family car. The pigeon succeeded somehow in traveling a hundred and twenty miles and locating the correct window, in the right building, in a strange town, at night and in a snowstorm.

Joseph Banks Rhine and his researchers at Duke University sifted through hundreds of cases of what they called "psi-trailing" in animals, trying to obtain precise verification. The one which most impressed them was that of a cream-colored Persian cat called Sugar who in 1951 seems to have trailed his owners across fifteen hundred miles of mountainous country between California and Oklahoma.[452] The family intended to take the cat with them, but it was afraid of cars and leaped from the window just as they were leaving Anderson at the northern end of the Sacramento Valley. They couldn't catch him again, but fourteen months later Sugar suddenly turned up, leaping through the window of their new home in Gage, Oklahoma. The cat had a deformity of the left hip which served as positive identification, easily recognizable by a veterinarian. How he crossed a desert, several canyons, and the entire width of the Rocky Mountains remains a mystery.

An even bigger mystery is why he should have gone to all this trouble in the first place. Why did the pigeon risk death to find the boy? What could possibly bring turtles and dolphins to the hands of human strangers?

It is pointless and circular to suggest that instinct is the answer. Instinctive urges are poor cloaks for ignorance. They were designed originally to describe observed behavior, then later offered as explanations for the behavior they were created to describe. It is equally dangerous and shortsighted to deny that unusual behavior patterns exist, merely because we cannot identify any physical stimulus that could have set them in motion. I have long felt that in our determination to give animal-behavior studies a solid scientific reputation free from the taints of subjectivity and anecdote, we have leaned over so far backward that we risk falling into a totally unnatural tangle. It has never been clear to me anyway why it should be regarded as objective to refuse to think about certain kinds of reality at all. Therefore it seems to me that we are forced to think at least in part in some sort of psychic terms.

This doesn't mean that we need end up talking lamely about loyalty and affection. These anthropomorphic emotions may well turn out to be appropriate stimuli for other mammals and birds and perhaps even for the occasional reptile, but it seems to be impossible to provide scientific proof for the existence of something like love, even in a human being. So the least we can do, which at this point also turns out to be the best we can manage, is to try to find a physical basis of some kind for the phenomena. There is no known electromagnetic link between a cat or a pigeon and its distant owner; and research into gravitational, radioactive, and nuclear reactions, though still in its infancy, has so far failed to show that any of these three remaining natural forces is involved in any way in apparent telepathic contact; so we are left with only one possible avenue of exploration. And all we can do is to look for similarities in the participants in the hope that these, rather than their differences, might provide some clue.

Everybody involved in all the examples I have offered—wasps and spiders, swallows and cuckoos, alligators and planthoppers, cats and pigeons, turtles and dolphins, mice and men—is a complex multicellular creature carrying its own distinctive genetic instructions, which, as far as we know, do not include do-it-yourself radio components. But all the communities in question also incorporate symbiotic microorganisms which retain their original characteristics whether they live now in an alligator's eye or a human salivary gland. We, all of us, share this identical inheritance. And I suggest, albeit a little diffidently, because my postulated "contingent system" is not much better than "instinctive urge" when it comes down to hard facts, that we look to shared patterns of awareness which in ourselves seem to exist almost entirely at unconscious levels.

And because it is necessary in this potentially hostile and almost totally uncharted country to grasp at the smallest straws, I see possible significance in the fact that the porpoise caller and the turtle toucher both talked about their contacts being made at the level of the dream. This proved to be a royal road for Freud, and it could be so for this quest as well. "Our dream," said Jules Renard, "dashes itself against the great mystery like a wasp against a window pane." Perhaps, but I believe it is worth persisting despite the fact that he added, "Less merciful than man, God never opens the window."[451]

# CHAPTER EIGHT

# ILLUSION: Personal Unconscious

In the crystal waters around Bermuda there is, among many others, a fish with the splendidly salacious name of slippery dick. *Halichoeres bivittatus* is a wrasse, about nine inches long, pale-green and yellow with a dark double stripe down its side. By day it is an active, even voracious part of the coral reef community, but at night it does something strangely unfishy. It sinks down to the bottom at dusk and buries itself in the sand, emerging again only at first light.[530]

On night dives around the islands I often sought out the resting wrasse for company, sweeping the sand gently away and lifting them carefully up from the bottom. If you do it without squeezing, you can examine the fish closely and even take flash pictures without producing any reaction. I once succeeded in lifting one unresisting wrasse completely clear of the water to show to a friend in a boat, before returning it to its ocean bed.

It is tempting to conclude that slippery dick actually sleeps, but since fish have no eyelids, it is difficult to distinguish between loss of consciousness and simple inactivity. Many terrestrial animals, even molluscs and insects, slow down to the point of standstill during the cool of the night, when their body warmth falls as fast as that of the air. But there is no difference between day and night temperatures in the waters of Bermuda.

Sleep, in its own way, is as hard to pin down as consciousness. We can describe sleep in man as a condition in which the eyelids close, the pupils become very small, the secretion of digestive juices and urine and saliva all fall sharply, the flow of air into the lungs diminishes, the heart slows down, and there are discernible changes of the electrical patterns in the brain. But these are merely the signs of sleep and tell us little about what it actually is. Despite the fact that we spend almost a third of our lives doing it, and a whole army of researchers devote all their waking hours to pursuing it, turning out more than six hundred scientific papers every year, we still know next to nothing about sleep, or why we do it.

Sleep seems to be a fundamental, built-in way of behaving.

It is not something we can normally choose to do or not do. It is an involuntary, but happily an adaptive, pattern of behavior; one that can be postponed in times of need, but not indefinitely. It has a habit of creeping up on most of us and forcefully reasserting itself again. Niko Tinbergen points out that sleep is a true instinctive pattern, because it is preceded by appetitive, or preliminary, behavior which includes all the symptoms of tiredness and the tendency to look for, or travel to, a special place.[543] It also involves particular and habitual postures, which may change as often as seventy times in a single night, but tend in general to be characteristic not only of individuals, but also of our species.

Sleep, like consciousness, is something we can recognize without difficulty in ourselves, but it is difficult to be certain about in others. All we can do is look for the telltale signs. Chameleons prepare for rest by curling up their tails in watchspring fashion and even close their conical eyes—independently, of course.[529] They don't, however, show any alteration of brain-wave patterns, so we can probably conclude that sleep, whatever it may be, appeared in its first full form somewhere between the reptiles and their warm-blooded descendants, the mammals and birds. There is no doubt that warm-blooded creatures sleep, and that many of them spend large chunks of their lives practicing this behavior. But why? And why, if it is such a widespread and important pattern, do the sleep requirements of different species vary so radically?

It seems reasonable to assume that sleep must have an important biological function. The easy answer, and still the most popular one, is that it has a restorative effect, allowing the body and the brain to refresh and renew themselves, synthesizing proteins and rebuilding nervous synapses. But there is little or no physiological evidence to prove that these things happen more often or more easily while we are asleep. Most of the facts usually used to support the recuperative theory are indirect ones, drawn from studies on the effects of not sleeping. In one study on goats kept awake for prolonged periods, it was found that they accumulated certain neurochemicals in their spinal fluids, and that these were capable of inducing premature sleep when injected into healthy, rested goats.[241] But this is evidence only that there are biochemical pressures enforcing sleep, not that sleep itself directly permits some vital physiological process which cannot occur during wakefulness. Drowsiness can be seen either as a manifestation of the need for the restorative powers of sleep, or, more simply, as one of the appetitive phases of instinctive sleep behavior.

The facts are that total sleep deprivation produces no change in the blood chemistry, no alteration of blood pressure, no measurable difference in heart rate or respiration, and no fluctuation in body temperature beyond the usual daily ranges.[574] Muscle action is not impaired and reaction time is not prolonged. There may, after several sleepless days, be some slurring of speech, loss of ability to focus the eyes, an inability to concentrate, episodes in which subjects lose all sense of time, a degree of memory loss, and hallucination, culminating even in severe psychological disturbance.[187] Ruthless interrogators have not failed to take advantage of these aberrations, but all of them can in the end be attributed to pressure exerted by the sleep urge. There is not a scrap of good evidence to show that our bodies as such must have sleep in order to function efficiently.[590] In fact, there are people who seem to be able to dispense with it altogether.

Several healthy insomniacs have been studied in sleep clinics and found to be perfectly normal, active, productive people, despite the fact that they seldom if ever sleep.[290] One seventy-year-old lady, who claimed never to have slept more than an hour a night, was kept awake and busy in a laboratory for three days and nights continuously, and finally slept for just ninety-nine minutes when she was eventually forced to lie down and rest.[372] Family histories suggest that this enviable ability is hereditary. In one case, both a man and his daughter were similarly afflicted, "much to the groggy consternation of his wife, who has since divorced him for a novel incompatibility."[471] It was he, of course, who got custody of the child.

Evidence from other species confirms that this wide variation in sleep requirement is not due simply to differences in metabolism. Warm-blooded animals whose surface area is large in proportion to their tiny size suffer heat-loss problems and as a result have high metabolic rates. The pulse of a short-tailed shrew *Blarina brevicauda* varies between 588 and 1,320 beats per minute. To maintain this level of activity, the shrew has to eat at least its own body weight each twenty-four hours, and to catch this many insects, it has to keep on going nonstop. Shrews apparently never sleep at all.[559] Bats of similar body size, however, deal with the same problem by coming to terms with the loss of heat. They simply cool down and let their metabolism dwindle almost to vanishing point, hanging cold and inert in the safety of their caves, sleeping as much as twenty hours every day.[370]

These divergent life styles produce radical differences in life span. No hyperactive shrew lives more than two or three years, but the bat, which takes its life in installments, survives for at least ten. It looks almost as though life can be measured in heartbeats, and it is up to each species to decide how best to space them out to gain the greatest benefits. Separate strategies, different daily rhythms, help to concentrate activity at those times of the day which have proved most profitable in the evolutionary history of each species. And where there have to be gaps in the program, these are filled by a new behavior pattern, the one least likely to squander valuable resources—by sleep.

This makes sleep a kind of ecological adaptation. Ray Meddis of London University calls sleep "the great immobilizer" and suggests that its prime function has nothing to do with restoration, but simply is to keep an animal still and quiet when it will be to its advantage to be so. "The critical feature of sleep control mechanisms is their ability to maintain immobility."[371] Not just to conserve energy but also to preserve life.

It is an appealing idea that sleep evolved to keep animals quiet when they were too young, too active, or not yet intelligent enough to manage this on their own initiative. If it is true, then sleep must be something you do only in your spare time. It can't be allowed to intrude into waking life where it would be counterproductive. And one would predict that those animals which need to feed or move a lot will sleep very little. The food value of grass and leaves is low, so herbivores like the elephant seldom doze for more than two or three hours in every twenty-four. While the swift and albatross, which need to keep aloft, and the dolphin, which must keep afloat, seldom if ever sleep at all.

The immobilization hypothesis makes much more sense than the restoration theory. It is particularly apt in the case of mammals, which it seems probably arose as warm-blooded creatures of the long cold nights, when dinosaurs ruled the days. It would have been absolutely essential for the early Mesozoic mammals to remain hidden during the time that predatory reptiles were active, and very difficult for them to do so in day temperatures that made them even more alert, unless they were immobilized in some forceful way. Sleep could have done this, and if it did play such a fundamental role in our early survival, it is hardly surprising that it still remains so deeply ingrained in most mammalian brains.

The most attractive feature of the ecological sleep idea, and the reason I have gone to this length to establish the natural history behind it, is that it provides us with a biological model that begins to make some sense of the nature and function of dreaming, and the role of the unconscious in our lives.

A small sleeping mammal, provided it can find a reasonably secure spot to spend its off-duty hours, has solved many of its problems. But it is still vulnerable, and a very high premium must have been placed on ways of even further reducing the risk of predation. The best bet in this situation would clearly be some sort of alarm system, which kept the animal immobile and yet allowed it also to be in a preparatory state of arousal. The dream does precisely that.

Our sleep is subdivided into cycles of alternation between two distinct patterns of activity. The first to occur is a synchronized rhythm of long slow brain waves, which accompany a general relaxation of most nervous activity. The eyes are still and the heartbeat is regular. Eighty-two percent of all mammalian sleep is of this kind, which has been called orthodox or slow-wave sleep, but is probably best described simply as "quiet sleep." Several times in each sleep session, this phase is interrupted by a much more volatile pattern which is characterized by short rapid brain waves, very much like those of the waking state, and by rapid eye movements behind closed lids, and an irregular heartbeat. This is paradoxical or rapid-eye-movement sleep. Since Eugene Aserinsky and Nathaniel Kleitman's discovery that the eye movements are almost always associated with dream activity, this phase of the sleep cycle has been known also as dream sleep.[23] But new evidence suggests that dreams may also occur independently of the eye movements, so to avoid confusion this pattern is now being called just "active sleep."

The active phase is so paradoxical and extraordinary that there is even doubt about whether it should be described as sleep at all. The developing fetus seems to spend nearly all of its time doing it. Having little or no outside information to go on, it stimulates and exercises itself. There is good evidence to suggest that it is these long periods of internally triggered excitation which are directly responsible for the rapid development of the more advanced mammalian brain and central nervous system.[456] So, for the fetus at least, active sleep patterns represent a sort of super-wakefulness, interrupted only by occasional periods of quiet sleep. This ratio persists for a while after birth, but as consciousness intrudes and environmental stimulation plays an increasingly large part in the infant's life, the proportion of active to quiet sleep decreases and finally inverts to the adult level in which it takes up less than twenty percent of sleep time, or about six percent of all experience.

So active or dream sleep is a time of mental arousal, a kind of autostimulation which keeps the system in trim, preparing the organism for physical arousal if this should prove to be necessary. It is usually accompanied by rapid eye movements, which may have something to do with dream content, and sometimes by slight twitching of the extremities; but there is also an ingenious nervous inhibition which simultaneously deactivates the skeletal muscles so that a dreamer doesn't actually thrash around making noises that might attract predators. This relaxation of the muscles amounts almost to temporary paralysis, so the nightmares in which we struggle to escape, but are unable to move, are a true reflection of our physical condition. Studies show that dreams of all kinds, and nightmares in particular, are usually directly followed by spontaneous physical arousal, with mind and body all set to go.[461] And because the active phase occurs four or five times each night, in man roughly every ninety minutes, sleep is economically divided up so it is guaranteed that we do, in effect, stop and check that all is well at regular intervals. Dreams are our sentinels in the night.

Of all living mammals the opossum *Didelphis marsupialis* is probably closest to those very first forms which we believe ventured out at night to challenge the supremacy of the ruling reptiles. Being warm-blooded, they had enough speed and agility to forage for insects and worms with special haste, and then scurry back to their hidden dens in caves and hollow trees, "waiting out the eons until their time came to inherit the earth."[495] These living fossils still sleep right through the day, spending about half their resting period involved in active sleep with rapid eye movements. Frederick Snyder of the National Institute of Mental Health in Maryland says that "the opossum is our witness that this extraordinary periodic central nervous system activation in the midst of sleep came to full flower at an early stage of mammalian evolution" and "was one of the crucial innovations which shaped the survival and ultimately the predominance of our kind."[495]

Studies in sleep laboratories confirm this vigilant function of dreams in man. Children, who are typically very difficult to arouse from sleep and normally become extremely confused if awakened suddenly, are alert if attempts to wake them are made only during active sleep.[196] Adults roused during quiet sleep show confused, repetitive thinking and even some loss of memory, but those awakened from a dream seem to come very quickly to terms with their environment.[174] And all subjects who are particularly anxious about the experimental situation wake fully and spontaneously right at the beginning of an active sleep period. Long dreams are luxuries we allow ourselves only in times and places of security.

Since dreaming begins, and occurs at such length, in an unborn mammal, we can conclude that its primary function is to stimulate and exercise the brain during its most critical period of growth. Later, because it is the most active part of a passive pattern of immobilization, dream sleep serves the secondary function of acting as a useful sentinel. But there is clearly a third function as well. Even the youngest human infants have much more complex visual and auditory awareness than we have given them credit for, and it seems likely that much of this flood of new information is incorporated into more or less ordered patterns during dreaming.[185] We have no right or reason to deny babies a rich dream life, nor can we exclude the possibility that most mammals experience vivid imagery and dream much as we do when instruments tell us that their eyes are moving rapidly and their brains are in a state of high excitement.

Snyder says that when he watches his opossums in the laboratory during periods of active sleep, their breathing sometimes becomes rapid and their limbs move as though they were running away from predators. This implies that they are hallucinating such a situation, which would in fact be a very good mental preparation for the real thing. An animal waking from such a state could carry the dream flight directly through into actual physical action almost without breaking stride. But Snyder also adds that "there are other occasions of assiduous licking or chewing, and this does not suggest appropriate preparation for an encounter with a hungry Tyrannosaurus."[495]

So, although preparation for danger seems to be the immediate and most valuable function of dreaming in the adult life of a primitive mammal, it doesn't end there. Sleep is still necessary for the safety of immobilization and the economy of energy, so dreams even in an opossum can also serve the function of fostering continuity of sleep by conjuring up images more inclined to gratify than frighten. And this would clearly reinforce the fetal function of mental exercise by giving the newly acquired mammalian cortex the kind of stimulation it needs to achieve its full psychic potential.

Freud said, "I do not myself know what animals dream of. But a proverb to which my attention was drawn by one of my students, does claim to know. 'What,' asks the proverb, 'do geese dream of?' and it replies: 'Of maize!!' The whole theory that dreams are wish fulfilments is contained in these two phrases."[188]

Freud, of course, went on to elaborate his theory into the complex structure of psychoanalysis, but in divining the integral relationship of dreaming to basic biological urges, he produced one of the most seminal scientific insights of our time. William Dement of Stanford University, one of the pioneers of dream research, says, "Here in a well defined area of the brain stem is housed a preemptive mechanism which initiates dramatic change in every corner of the nervous system. In its task of modulating the activity of the entire brain, it has at its call widely disseminated neural networks and highly specialized biochemical and physiological processes. If such a strange and complex phenomenon evolved in mammals for no reason at all, we must doubt some of the most basic propositions of the biological sciences." And he adds that "never before in the history of biological research has so much been known about something from a descriptive point of view, with so little known at the same time about its function."[127]

The theory of natural selection demands that behavior be adaptive. And if a behavior pattern, such as active sleep, is widespread through many different kinds of animal, it must in some way be important for all of them. With the new suggestion of a sentinel value to add to the descriptive information already accumulated, I believe we now have almost everything necessary for a full evolutionary picture of the value and function of sleep and dreams. This is my scenario:

In the beginning, life was little more than a chemical reaction. All such reactions depend on heat, in general taking place more quickly as the temperature rises, and occurring more slowly as it drops. If there ever was a primeval soup, and it was large enough to be relatively independent of diurnal temperature changes, all life would have carried on as shrews do today, without pause. In the depths of ocean trenches where temperature, illumination, and pressure never change, this is probably what still happens. But in the surface layers of the ocean, in shallow pools and on the land, where circadian patterns prevail, almost every living process comes to take on a typical diurnal rhythm. Even plankton rises and falls every twenty-four hours. This is the natural rhythm of our planet, and very few organisms can avoid it. Hardly any bother even to try, because there is survival value in having alternating quiet and busy periods. For most it is true that a pause refreshes and prolongs active life. So the first great rhythm in life was one which simply distinguished between activity and inactivity.

Most invertebrates are still stuck in this phase, but the most complex ones face a new problem. A sponge can just switch off, shutting down all the cells in its community at the same time, because they depend on each other mainly for the function of feeding. When this stops they still stick together, but to a great extent they cease to be a community in any real sense, and become a heap of inactive cells that just happen to have come to rest in the same place. But in more complex associations, there are community functions that can't be closed down. Vital services can be reduced, but still have to be maintained. Simple hearts must carry on beating, nourishing tissues that have become too specialized to cater for themselves. Rudimentary nerve networks must continue to broadcast at least a simple program designed to keep certain activities just ticking over. If these systems stop altogether, the community loses its identity and ceases to be an organism that collectively bounces back into action when the next period of activity begins. So at quite an early evolutionary stage, there was already a need for two separate nervous systems: one to cope with fluctuating daily needs, and another to carry on regardless.

This nervous discrimination reaches its highest level of sophistication in vertebrate animals, which have, as we have seen, a very special and highly ordered existence. All animals with backbones have a separate and distinct autonomic nervous system, which supplies the gut, the blood vessels, and the glands. And it is this which governs all the unthinking, automatic services which keep our communities running smoothly during periods of inactivity. In cold-blooded vertebrates such as fish, amphibians, and reptiles, the autonomic system predominates, but a great deal changed with the advent of temperature control by the warm-blooded mammals and birds.

Internal temperature control eliminates the major environmental variable. With it, all the basic chemical reactions of life can carry on independently of outside fluctuations. The organism is free to behave with far more variety. It has a bigger personal say in when and how things can be done. It becomes, in a very fundamental way, a free agent, a person in its own right and for the first time. It can establish full individual identity. The first and most direct consequence of this emancipation was that a new central nervous center developed to coordinate the more complex behavior patterns. This was the limbic system.

An ant is just a ganglion on legs, a little solid mass of nervous tissue near the head with offshoots to the relevant appendages. In reptiles there are similar spidery systems scattered through the peripheral network of the skin, but all these outlying areas, like a host of sensory insects, are under the central control of a different kind of master, a nerve complex protected by the spinal cord. This swells somewhat at the head end largely because it is here that most of the major sense organs are localized. Not much more than this was necessary, but when mammals came to be, an extra layer had to be added to cope with new demands, new feelings. This took the form of a limbic knob surrounding the head end of the spinal cord, almost as though the old reptilian brain complex had been decorated with a new and more fashionable hat.

The mammalian limbic system seems to be concerned mainly with emotions, something no organism ever had to cope with before. As Carl Sagan puts it, "the reptilian mind is not characterized by powerful passions and wrenching contradictions but rather by a dutiful and stolid acquiescence to whatever its genes dictate."[471] But in mammals, things are very different. Electrical stimulation of one little almond-shaped area of the limbic region can rouse its owner to extraordinary levels of frenzy. In one experiment a cat wired up in this way could be induced to cower in fear at the sight of a mouse. In another, an aggressive lynx was calmed to the point where it could be patted.[485] Love, hate, fear, anger, some sex, and much parental care seem all to be organized here.

Many of these emotions involve autonomic patterns; blood vessels dilate, hormones are secreted, bronchi expand, hair stands on end, eyelids retract, pupils change in size, and the liver goes into overdrive to feed muscles involved in things like fights and flights. So in animals which are warm-blooded (even the term itself is linked with the ability to display emotion), there has to be a center which coordinates activity between the voluntary and involuntary nervous systems. This too is part of the limbic region and accounts for the ability of many species to bring the smooth muscles of internal organs under conscious control. The bushbaby *Galago senegalensis* urinates on its hands just before each jump, giving it both a better grip and an easy way of marking out its territory with distinctive smells.[153] Howler monkeys *Alouatta palliata* in the canopy of South American forests defecate directly and accurately down on intruders on the ground below.[100] The coordinating center is also responsible for psychosomatic illnesses such as ulcers, providing direct links between stressful external conditions and the equilibrium of internal states.

All this is well known, but perhaps the most far-reaching consequence of the limbic bridge between active and inactive nervous systems is that it establishes a sort of schizophysiology. The two systems have partly overlapping functions, and neither had clear-cut dominance over the other. This was the situation for the first mammals, and it is one in which the opossum still finds itself somewhat embroiled: part reptile, part mammal, and a good part alimentary canal. I believe this conflict eventually provided the stimulus for a major reassessment of identity, and that it produced the impetus necessary for the rapid and revolutionary development of the cortex. But the first result was that it invented properly coordinated sleep.

The first warm-blooded birds and mammals arose from independent reptilian sources at roughly the same time. Because birds could fly, they were able to keep out of the way and yet remain in the day, while the mammals were virtually forced to take to the night and to hide as best they could during the dangerous daylight hours. When the dinosaurs were active, mammals had to be inactive, and therefore largely under the control of the autonomic nervous system, which normally sets up a regular pattern of stimulation to keep internal mechanisms idling at the right speed. I suggest it was this metronomic signal which combined with the reptilian brain to produce the synchronized pattern of brain waves, and the characteristic inert behavior, which we now recognize as sleep. Sleep is inactivity plus coordination.

Another result of warm-bloodedness for mammals was an ability to dispense with eggshells and allow the embryo to grow instead to relative maturity in the constant environment of the uterus. Here it recapitulates its evolutionary history, running through all the patterns of the past and ending the sequence with a triumphant purely mammalian flourish by pushing its cortex on to new heights of integration and initiative. To achieve this it draws on the stimuli already present in the old reptilian brain and elaborates them into a cortical storm which rages almost without pause through the final embryonic stages, incorporating even rapid eye movements. The result of this electrical activity is known to be a stimulation of growth and coordination,[456] but little thought seems to have been given to its immediate effects.

If an eight-month-old human fetus has an excited cortex, it is being stimulated in the frontal lobes, where deliberation will later take place; in the parietal lobes, which are concerned with spatial perception; in the temporal lobes, which are designed to handle a variety of complex tasks; and in the occipital lobes, where vision will eventually be coordinated. But it has no visual information to process, it knows little of space beyond the confines of the uterus, and has nothing to think about except the regular rhythm of its mother's heart and a confusion of womb-filtered sound. This may be more than enough to keep it busy through the long hours of active sleep, but I suspect not.

I offer instead the suggestion that it is at this time, and in this extraordinary superwakefulness, that an individual makes first contact with the contingent system. That it is here, in concert with a physical reiteration of phylogenetic stages, that it also experiences a mental playback. It hears, in a very simple, rather diffuse, timeless, tacit way, the voice of nature. It is exposed to, and at least partially programmed by, racial memories and archetypal forms. The fetus learns first to recognize the music of earth, and only later concerns itself with the problems of becoming an individual human being.

I know this is frivolous and mystical and totally unsupported by scientific fact, and yet there is this matter of the fetus with much on its mind. It is time we at least began to consider possibilities. This is one which could begin to provide a foundation for Jung's collective unconscious. At the very least it might help explain some things we do, some things we all seem to know, without having the necessary genetic programs. This could be a way to pass on after conception, but before birth, information that those around us have only recently acquired. It could vindicate Lamarck.

So much is outrageously speculative, but one thing is certain. When the first mammals needed an alarm system to stand sentinel over their newfound capacity to sleep, it was waiting there ready-made in the embryonic brainstorm. The ability to dream was already part of mammalian inheritance and needed only to be incorporated in quiet sleep as an intermittent active phase.

The most effective sentinels were nightmares, dreams of dragons which left the sleeper poised for appropriate action. Then the reptiles disappeared and mammals began to eat one another instead. The prey dreamed less, in snatches and in shifts, and the predators dreamed more. The most successful predators had the greatest time to sleep and dream, and the least to fear from doing so. Given this freedom, there was room for some of the dreams to be good. And the longer and more involved the dreams became, the more they exercised the growing brain, and the better it came to be. It is easy to see how, with expansion both of the capacity to dream and the time to do it in, the dreams must have come to exercise selective pressures of their own. They probably started with simple wish-fulfillment of the geese-and-grain variety, but must soon have begun to feed voraciously on all available information, drawing on personal experience and the growing fund of emotional response. One of the theories of dream function suggests that it helps to process experience, ordering events and deciding which to dump, and which to place in the stores of long-term memory. It is certainly true that people involved in demanding intellectual activity during their days require more sleep and a greater chance to dream than those whose waking time is spent in repetitive, less challenging tasks.[241] Dreams now seem to serve memory in this way, but they might also have been responsible for creating an effective memory in the first place.

Freud set up his personal unconscious as a shorthand term for a complex set of events connected with unpleasant memories that had been repressed, but which he believed were nevertheless still powerful enough to instigate dreaming as a way of working off tensions. And he suggested that this was a sort of safety valve that helped not only to preserve sleep, but also to reduce neurotic symptoms during wakefulness. The discovery of active sleep as well as quiet sleep throws doubt on the unconscious as an instigator of dreams, and it tends to be seen now by psychoanalysts more as "a prepotent force" which is likely to invade ready-waiting dream sequences.[345]

Freud also thought that our tendency to forget dreams was a direct result of the action of a censor mechanism. Now that dream-laboratory studies have shown how often and how long and in what fine detail we dream, it seems unlikely that a censor could object to all the multitude of information that is included. But though he may have been partly proved wrong about the mechanism, there is no doubt that many dreams still express the content of the unconscious very strongly, and nobody has ever improved on Freud's method of dream interpretation as a "royal road" to an examination and an understanding of the personal unconscious.[244]

To a biologist, one of the most interesting constructs to emerge from Freud's distinction between conscious and unconscious processes is the concept of an ego. It is seen as an organized part of the psychic apparatus. In Freud's own words, "The ego is first and foremost a body ego, i.e. the ego is ultimately derived from bodily sensations."[191] It is in effect the image that the self has of its own physical system. A baby's first grasp of our objective reality is a grasp of his physical self, of his existence as a separate entity with distinct limitations and less-easily defined possibilities. And as it grows, it sustains this body image by continuous stimulation. Its notion of itself is constantly reinforced by feedback from the environment. This corresponds very well with my suggestion that sensitivity in a developing organism is what leads, given sufficient complexity and time, to identity, to self-awareness.

Barbara Lerner of the Mental Health Center in Chicago points out that, to start with, a human child's needs and desires are comparatively simple, and in the normal course of events, are easily and directly fulfilled. It experiences its environment in ways that tend to reinforce its body image. But as we become older, desires are not all acted out. Some are merely delayed, while others are denied altogether, which produces a gap between soma and psyche, between the body and the mind, and that this results in a considerable weakening of the body image and the ego. She suggests that an important part of a dream's function could be to reintegrate the personality by allowing us all to act out our fantasies.[343]

Recent research on cats at the Harvard Medical School has finally proved that rapid eye movements are directly produced by signals from the old reptilian area in the brain stem, which contains giant nerve cells that fire in sequence, sending signals to the new mammalian forebrain, where these result in an explosion of electrical activity in the cortex, producing, among other things, the movements of the eyes. Allan Hobson and Robert McCarley say that "it is amazing that we dream anything more than a kaleidoscope of images and colors—that we see anything orderly at all."[20] But if Barbara Lerner is right, then rapid eye movements may be something like Rorschach ink blots and have no meaning anyway, merely providing an unstructured visual stimulus which the dreamer then shapes into meaningful configurations that usually involve imagined action. Perhaps just the sort of action that will best satisfy the repressed fantasies, and most effectively reinforce and re-establish a fading body image and a shaky ego.

Now that we know about the strangers in our midst, and about the relative independence of consciousness from genetic and contingent systems, it becomes more important than ever that we have some way of holding the whole circus together. It may be only in dreams that conscious thought, voluntary sensation, unconscious reaction, involuntary physiology, genetic pressure, and the separate programs of subcellular organelles can all meet in committee to sort out conflicting needs and rival demands. Agreements reached here, probably on a time-sharing basis, may be all that keeps our complex concerns in business at all.

Paul MacLean of the National Institute of Mental Health in Maryland sees the conflict in man purely as one between the archaic structures of the brain and the new specifically human neocortex. They have certainly been superimposed on each other as a result of rapid leaps in evolution, apparently without the precaution of ensuring adequate coordination. He says, "Man finds himself in the predicament that Nature has endowed him essentially with three brains, which despite great differences in structure, must function together and communicate with one another. The oldest of the brains is basically reptilian. The second has been inherited from the lower mammals, and the third is a late mammalian development." And he suggests that "speaking allegorically of these three brains within a brain, we might imagine that when the psychiatrist bids the patient to lie on the couch, he is asking him to stretch out alongside a horse and a crocodile."[398] Or an opossum and dinosaur.

I suggest that such conflict is not merely a human concern, but exists in all mammals. The opossum has such trouble coming to terms with its reptilian ancestry that it ends up sleeping for as much as twenty hours every day, and dreaming for ten of these. The tensions are real and may in man, with his triple troubles, have resulted in mental imbalance. Arthur Koestler sees us as torn; "on one side, the pale cast of rational thought, of logic suspended on a thin thread all too easily broken; on the other, the native fury of passionately held irrational beliefs, reflected in the holocausts of past and present history."[322] There is no denying our dilemma, but I believe MacLean and Koestler, in their real concern for the present, are not seeing the problem in its true evolutionary perspective. We wouldn't have our shiny new cortex at all if it hadn't been for comparable tensions in our early mammalian development. Natural selection feeds on pressures which produce instability and a demand for change. The tensions are essential; we couldn't have become human without them. And though we might be now in dire straits, and may even destroy our planet entirely, there is no knowing what extraordinary results a resolution of these tensions could bring. A quadruple brain perhaps, with even greater problems coupled with unimaginable strengths and skills? Or a quantum jump in some totally new direction?

I believe in fact that the triune brain is only part of the problem. We are, as I have been trying to demonstrate, subdivided in even more fundamental ways into a variety of evolutionary stages, each with its own peculiar form of subjectivity and its own intelligence, its own sense of time and space, and many of its own memories. And the only one among us to sense this conflict in all its dimensions seems to have been Carl Gustav Jung.

For Jung, as for Freud, the dream was the principal instrument of therapy. It was the one psychic phenomenon that permitted relatively easy access to the contents of the unconscious. But Jung differed from his mentor in that he saw in dreams not only the signs of personal conflict, but something more, something which seemed to go back to a primordial experience, to universal problems. He refused to consider any dream situation out of context and questioned the assumption that there were any standard symbols that could be translated, as if out of a dictionary. His most valuable insight, for biology at least, was that there was no such a thing as a dream unit. He said, "It is on the whole probable that we continually dream, but consciousness makes while waking such a noise that we do not hear it." This is a superb idea which makes such perfect biological sense that it can't help but be true. It makes dreaming a purely natural phenomenon, an autonomous process with its own language and laws. "One does not dream;" said Jung, "one is dreamed. We undergo the dream, we are the objects."[282]

There is still room in this process for personal intrusions. Remnants of the day's events, impressions from the day before, memories of other similar occurrences, all form part of "the constellated contents of the unconscious" and creep into the continuing monologue. But they are so fragmentary that a full picture can only be gained by looking at them over a period of time. Jung was the first to insist on the importance of a whole dream series in analysis. And he tried in this way to enter into what he called "the land of childhood" where the rational consciousness of the present was not yet separated from "the historical soul." He realized that an individual human's drift away from this source was inevitable, but that it resulted nevertheless in disorientation. "The resistance of consciousness against the unconscious, as well as the underestimation of the latter, is an historical necessity in evolution, for else consciousness would never have been able to differentiate itself from the unconscious at all."[296] We needed the tension of the Lifetide, but that doesn't necessarily make it any easier to live with now.

All dreams tend to have their own momentum and intelligence, operating free from our usual constraints, as if they did indeed simply pick us up and carry us along as objects being dreamed. They ignore conventional space and time, and occasionally may even leave us with knowledge or information that seems to be from another space or time.

Telepathic and clairvoyant dreams are rare, but they do occur. Nebuchadnezzar, Alexander the Great, Charles Dickens, and many others have had them spontaneously. Some, like the biblical prophet Daniel and the modern "sleeping prophet" Edgar Cayce, actively sought them out and seemed to have managed to find them.[104] Freud, despite his antipathy to the collective unconscious, once said that "it is an incontestable fact that sleep creates favourable conditions for telepathy."[190] Most of the evidence remained anecdotal until Montague Ullman and Stanley Krippner set up their dream-research laboratory at the Maimonides Medical Center in New York in 1962. In ten years of intensive study they have succeeded in producing enough evidence to leave little doubt of the reality of telepathic dreaming, and a few experimental results that are strongly suggestive of clairvoyant dreaming.[552] But to me the most significant aspect of this work is the fact that, despite their vigilance and the elegance of their technique, the successful tests form a minute part of their total effort.

When someone does have a telepathic or premonitory dream which gives them information about, or even helps them to avoid, a crisis—it is news. Cases of this sort do occur, and when they do turn up they are very properly regarded as extraordinary.[219] The vast majority of disasters, misfortunes, and accidental deaths still occur without warning and go unnoticed until news about them arrives in the usual ways. Most of the information included in dreams, and in all other forms of apparently psychic communication, is of a trivial nature. It has little or no adaptive value, at least as far as the individual recipient is concerned. Jule Eisenbud of the University of Colorado suggests that information at the dream level "is geared primarily not to the individual but to an interlacing hierarchy of eco-systems in which the individual, regardless of species, is necessarily embedded."[158] It does not establish any useful contact between individuals, or restore personal equilibrium, or enhance self-awareness, or provide constructive answers to particular environmental problems. Nor is it surprising that this should be so. One learns little about honey from an isolated bee. If, as Jung suggests, the dream is based primarily on a collective unconscious which is connected to all living things, then we should expect it to have more universal concerns.

Maurice Maeterlinck, the Belgian poet and scientist, worried about "the strange, inconsistent, whimsical and disconcerting character of the unknown entity within us that seems to live on nothing but nondescript fare borrowed from worlds to which our intelligence as yet has no access. It lives under our reason, in a sort of invisible, and perhaps external palace, whose interests, ideas, habits, passions have naught in common with ours. Is it making fun of us?"[357] I doubt it. There have been times in my exploration of the supernatural when I have been tempted to conclude that the whole thing is some absurd cosmic joke. So often things happen without reason, in ways that seem childish or even absentminded, usually without any redeeming practical value. But I sense that it is wrong to judge these events by our necessarily restrictive standards; we have to try to fit everything into a truly ecological perspective, or otherwise we end up like the hapless detective Inspector Clouseau, picking up nothing but our own fingerprints.

If Jung is right about an ongoing natural process which our personal unconscious samples four or five times each night in dreams, then there might well be other ways we can experience this flow. It may be possible for us to tap into it consciously, like the poet in *A Midsummer Night's Dream* who takes the forms of things unknown and turns them into shapes "and gives to airy nothing a local habitation and a name."

Hieronymus Bosch and William Blake did it visually; Samuel Taylor Coleridge and James Joyce managed it with words. They succeeded in diverting the stream of consciousness in ways that allowed dream imagery to survive in the harsh light of day. They were all exceptionally gifted men, but each one of us has the capacity to transform experience in equally magical ways. We do it every time we daydream.

Memory studies show that we store an alarming amount of material, processing it through the brain and filing it away in some still mysterious location. Nobody knows quite where this is, but we are beginning to learn a little of how it works. Our information storage center is nothing like the static data center in a computer; it is an active system in which much of the material is kept in a state of constant upheaval. A stream of thought flows endlessly through, stirring things up and keeping some memories floating up near the surface where they can most easily be retrieved. We have, in effect, direct access to a vast pool of images and associations, all ready and waiting for rescue. Normally we use very little of this reservoir, just enough to meet our daily needs, but every now and then, without any apparent reason, the mind takes a dip at random.

We call it reverie, woolgathering, making pictures in the mind's eye, or building castles in Spain. Everybody does it, least often when we are involved in something supersensory like eating, and most often just before falling asleep. But it is a process quite different from quiet or active sleep; it shows nothing like the ferment of a dream. There are no known physiological symptoms of daydreaming. Pulse rate is unaltered, respiration continues normally, nothing dramatic happens to the brain waves or the electrical resistance of the skin. There is, however, one way you can tell if someone is doing it: The gaze becomes fixed straight ahead, and the movement of the eyes slows down and stops altogether as they drift gently out of focus.[489]

Gaps between episodes in a daydream are marked by what Jerome Singer of Yale University calls "the concertgoer's effect."[21] Just as everyone at a concert coughs and shuffles between movements, so the daydreamer's eyes come back to life and they swallow or even sigh as the spell is broken. It seems that in order to daydream when wide awake, we need to suppress information coming in from the environment, and because we are predominantly visual creatures, that means canceling out the eyes. Not by closing them, because this leads to physiological changes associated with sleeping, or leads to compensatory effects in which the ears become even more finely tuned, but by a sort of self-imposed visual monotony. Similar effects can be achieved by sensory isolation in black rooms or with broadcasts of white noise, but most of us, it seems, can isolate ourselves even in a crowd without resort to complicated equipment. John Antrobus and his associates at Yale University have confirmed this effect by putting subjects in a situation where their visual environment consists of nothing but a screen on which moving vertical stripes are projected.[22] This leads to an optokinetic reflex in which the eyes automatically track up or down the screen in the same direction as the movement. Whenever the eyes stopped, the subjects were asked what they were thinking about, and always it turned out to be some elaborate visual fantasy.

Active, objective thought is always accompanied by ocular activity. It is even possible to tell what kind of thoughts are taking place by the direction in which the eyes move. Paul Bakan at Michigan State University has discovered that shifts to the left are indicative or emotional, aesthetic judgment and visual imagery, while eyes that move to the right betray a more logical, analytic train of thought.[28] Over long periods of time artists, who are, it seems, people with dominant right cerebral hemispheres, will tend to favor left shifts, and scientists, or those who are mainly verbal left-brain users, move most often to the right.

Many daydreams are of a Walter Mitty, wish-fulfillment kind. Psychoanalytic thinking, which until very recently thought all imagination grew out of suppressed desires, has tended to classify them as neurotic and escapist. In most cases daydreams are described as "regressive" and "immature." But for biologists, that kind of judgment is nonsensical. Behavior as widespread and clearly defined as this must, we believe, have adaptive significance. It has been clearly established in studies on sensory deprivation that what holds us together functioning in an effective way is not the amount of external stimulation, but its variety.[173] When one's environment is limited either in stimulation or in variety of content, one usually begins to shift attention inward. A monotonous task in an unchanging environment on an assembly line leads almost inevitably to mishaps, but not because of reverie. Analysis of industrial accidents shows that most occur because workers are distracted or lose concentration as a result of joking or teasing with others, in desperate extrovert attempts to provide external variety. Programmed music or pleasant and evocative smells, on the other hand, are more likely to lead to introspection, and the evidence shows that daydreaming actually helps keep workers alert during otherwise monotonous tasks.[489] A self-induced pattern of activity in the right brain balances up, it seems, with a predominantly left-hemispheric task, and makes it less likely that this will be distracted by competitive analytic programs. We can create our own musical accompaniment.

Viewed in this light, the daydream has survival value. Many of the tasks involved in maintaining life are routine and repetitive. Our burgeoning brains dislike them and try to avoid them wherever possible. The most basic retreat for waking consciousness is to become less conscious; we tend to react to boredom by falling asleep. But often this is totally inappropriate, and could even be lethal. Hence the daydream. Like the night dream, which stands watch over sleep, keeping the system in a state of preparedness, the daydream punctuates our waking state with interludes of mental activity designed to keep us ticking over continuously. I suspect that, like night dreams, daydreams occur at more or less regular intervals throughout our waking hours, perhaps following the cyclic patterns we have now begun to identify as biorhythms.

There is one point where day and night dreams almost meet. Many people start to sleep, or at least their brains show the long slow delta waves characteristic of quiet sleep, with their eyes half open. And they seem in this state, even though they show none of the eye movements of active sleep, to be dreaming. When roused, they report complex visual images and associations. David Foulkes at the University of Chicago has found that this happens most often with subjects who are given to introspection, who show less anxiety and more psychological self-awareness.[181] He interprets these features as indicative of those who are most liable to dream immediately on falling asleep. But they are precisely the traits most characteristic of good daydreamers. And it seems to me that what he has been recording is not so much displaced night dreams, occurring unusually early in sleep, but untimely daydreams, taking place a little later than usual.

I think the answer is that there is far more permanent associative activity going on in the brain than we appreciate. I suggest that consciousness is a continuous process and doesn't simply stop when we are asleep. Even enforced loss of consciousness induced by chemical anesthetics doesn't necessarily interrupt the flow, as surgeons have discovered to their cost when patients answer back from the table in the operating theater, or repeat conversations they were never meant to hear.[344]

Sleep is therefore not a suspension of consciousness, but a continuation of it under slightly different control. And a dream, as Foulkes puts it, "does not burst like sudden firework against a background of complete darkness; it develops in a context of already ongoing mental activity."[180] Dreaming seems to be a continuous process in its own right, carrying on parallel to consciousness, and it is probably quite wrong to assume, as some have, that it is an altered state of consciousness. As a process it precedes consciousness in the development of the embryo, where it dominates psychic life. And it continues to do this throughout our lives at all times when consciousness is dulled, as it necessarily is during sleeep. Unitary dreams are, I suggest, moments when the dream stream and the flow of consciousness overlap and intermingle. The fact that two distinct forms are involved is, I think, reflected in the existence of a whole series of mixtures that can be graded from pure dream to pure conscious stream.

At the dream end of the scale are exhilarating flights of total fantasy in which all disbelief is suspended and anything can, and usually does, happen. Then there are the usual complex combinations of reality and mystery, of personal and collective concerns, with which most dreams deal. This is naturally the happiest hunting ground for psychoanalysis.

At the other end of the spectrum, with increasing consciousness, are the pre-lucid dream, in which the subject wonders whether or not he is dreaming or awake, and may or may not come to the correct conclusion; and the completely lucid dream, in which the subject is aware of dreaming and may even decide to influence the content of, or the action in, the dream. It is significant, I think, that lucid dreams, in which the threat of waking is great, are accompanied by the most comprehensive action of the mechanism which suppresses body movement and muscle activity, almost as though the control center in the brain stem recognizes the risk of disturbing sleep and goes all out to minimize it.

The rare nightmare is a special case, occurring most often during the latter part of the night, and nearly always at the end of a long dream cycle.[242] I suggest that it is probably an emergency system, a sudden forceful and traumatic intrusion of an aggressive consciousness into the dream, in which we literally frighten ourselves into wakefulness. It looks like a mechanism designed to keep the dream stream from taking over altogether. And it is most common in young children simply because they are still largely under the embryonic thrall of the dream stream and in the process of establishing a working relationship between it and growing consciousness.

When we are awake, the tables are turned and consciousness is dominant. But I suggest that there are nevertheless a number of ways in which the deep currents of the dream stream manage to surface even in the bright light of day. And the daydream is only one of these haunting experiences.

# CHAPTER NINE

# DELUSION: Collective Unconscious

Some years ago, on my way back from one of the wilder parts of eastern Asia, I paused to take stock in Singapore. Living is easy there and the food is good. It is a splendid place to recover from malaria and dysentery, and I planned simply to regroup rather than pursue any particular project. But I had forgotten that in addition to being a giant and very efficient cash register, Singapore is also the world's most fascinating anthropology class.

It was January, the month of Thai, and the star Pusam was in the ascendant. The monsoon season was almost over, and in the Hindu temples of Chettiar and Sri Mariamman, the faithful were making preparations for a festival dedicated to the Lord Subramaniam.

Visitors are always welcome in Tamil temples, and I spent many lazy hours at Chettiar, sitting barefoot on the cool stone floor next to a glazed cement sacred cow, chatting with a friendly priest between bells. Every time the bells began he would leave for the inner shrine and reappear with the sacred white ashes of Siva, to pass among the congregation, bearing blessings. I was much blessed in those peaceful days and more than ready when the celebration of Thaipusam drew near to take an active part in the ritual.

For three days and nights I lived in the temple precinct, sleeping little, eating nothing and drinking only holy water, until there was no rhythm in my life but that provided by the ecstatic beat of the *molam* drum and the wail of the sacred trumpet. On the first of the two days of Thaipusam we went in small groups up to Bukit Larangan—the "Forbidden Hill"—and came back down in procession praising Subramaniam, asking forgiveness and offering thanks. Many of the faithful mortify themselves, pushing steel spikes through their lips and cheeks, suspending pots of milk from fishhooks barbed deep into the flesh on chest and back, or staggering along for miles impaled on the points of steel spears in cages weighing up to seventy pounds. Finally, in spasms of religious abandon, they collapse on the temple stairs, where the priests treat their wounds with ash and lemon juice.

My participation was more modest. I managed on the first day to limit myself to a single green lime, the symbol of purification, suspended on a silk thread from a pin pushed through the relatively insensitive tissue of my left earlobe. But by dawn on the second day, my fifth without food, I was so entranced by thick clouds of incense and the heady scent of frangipani flowers that I agreed to take part in a firewalk. I would have crawled through the flames to get a decent meal.

A cinder pit about sixty feet long was prepared in the yard of one of the smaller temples, and once again we came down through the town in procession, starting up in the hills at dawn and dancing barefoot all the way. We came into the temple in pairs, with an old Tamil from Madras wearing nothing but a white cloth around his waist holding me by the hand. Three times we circled a tower of glossy multicolored statuary reputed to house a pure-white cobra visible only to the pure in heart, and each time we passed directly through the embers smoldering in the pit. I wasn't the least bit concerned about burning my feet, but remember on the last circuit wondering idly why my cotton trousers didn't catch fire.

I recall very little else of the remainder of the festival, except that I slept long and well and woke only to the sound of drums on the following day to find myself on a straw mat in a corner of Chettiar cloister, feeling marvelous. I went down to the octagonal market at Telok Ayer and ate a huge meal of Hainnanese rice and an assortment of Malay fish curries. And then, still oozing well-being, I went along to sit on the banks of the Singapore River watching the acrobatic maneuvers of the watermen on the *tongkang* lighters in the overcrowded pool. And here my equanimity was really shattered.

I was joined by a friend, a photographer on *The Straits Times*, who had spent two hectic days on special assignment— a photo essay for a magazine on Thaipusam. With the crowds at the festival I wasn't surprised that I hadn't seen him in action, and I expected a little gentle teasing about my role as the only Western participant in the firewalk. He enthused about the pictures he had taken in the pit and was sure these would prove to be the highlight of his story, perhaps even suitable for a cover. Finally, unable to contain my bursting pride and curiosity, I asked him if he had taken any of my crossing of the coals. He simply looked blank and mystified, and proceeded to heap mystery on my head by insisting that he had taken up a good position on the edge of the pit when the flames were first lit, and not left there until it was all over. He hadn't seen me at all.

Later we went together through all his pictures and even found my old Tamil friend, but there was nobody holding his hand. I wasn't there.

I spent the remainder of the day in some mental distress, wondering whether the malaria had finally got to my brain. Established boundaries seemed to be collapsing all around me, leaving nothing firm enough to give me footing. I wasn't sure any longer that there was any concrete reality at all, and I was afraid to go back to the temple to find out. I am still unsure exactly what happened during the gap in time of Pusam, but when I returned to my little Chinese hotel that evening to have my first proper wash in days, in the shower room up on the roof where the proprietor grows his orchids, I discovered something that compounded the mystery, but made me feel a great deal more secure. The hair on both my legs around my ankles and for several inches up each calf was singed.

In 1756, Johann Leidenfrost, a German physician, placed drops of water into a spoon heated in a fireplace and noted how long they lasted by counting the swings of a pendulum. He discovered that the hotter the spoon, the longer each drop survived. This apparently contradictory phenomenon is well known to pastry cooks, who traditionally heat their stoves until water sprinkled on the surface remains there for a minute or more, dancing and skittering on the metal, instead of evaporating in a few seconds. But it wasn't until the original Latin paper was translated and republished in 1966 that physics began to take a real interest in the Leidenfrost Effect.[342]

If a surface is sufficiently hot, the bottom of a falling water drop vaporizes as it nears the metal, leaving a layer of vapor to support the remaining portion about one-tenth of a millimeter above the surface. Here it hovers, reasonably safe from the heat, until conduction vaporizes the remainder of the drop. The temperature at which the drops last longest is now known as the Leidenfrost point for that liquid.[213]

The American physicist Jearl Walker demonstrates the Leidenfrost Effect by wetting his fingers and then plunging them into molten lead at five hundred degrees Centigrade. Meat cooks at one hundred degrees, but the wet fingers are protected, for a few seconds anyway, by a sheath of water vapor. And he suggests that this could be the secret of fire-walking. As each step places parts of a foot in contact with the coals, the natural moisture vaporizes partially to give momentary protection, and, provided the steps are not too fast, sweating between footfalls could be sufficient to replenish the moisture. Which could explain why firewalkers, whether they perform in Fiji, Indonesia, or Malaysia, tend to live in humid tropical areas and all emphasize a deliberate ritual rhythm. And why, in Singapore at least, the participants were encouraged to drink lots of holy water.

Walker had such faith in his physics that he even went to the lengths of testing it himself in a five-foot-long bed of white-hot coals. "I suddenly found it remarkably easy to believe in physics when it is on paper but remarkably hard to believe it when the safety of one's own feet is at stake. As a matter of fact, walking on hot coals would be such a supreme test of one's true belief in what one had learned that I have suggested graduate schools might substitute it for the Ph.D. examination in physics."[567] He passed with honors and without harm.

I would like to think that I too had earned my laurels the hot way, but in a sense my experience in Singapore is even more interesting. Firewalking has been performed often enough to be beyond dispute, but the whole question of hallucination remains very much open.

Experimental studies are rare. At Yale, volunteer subjects were asked to walk down a long corridor until a light flashed.[480] There was in fact no light at all, but more than half those tested stopped anyway. Another study involved the use of an impressive piece of apparatus which, it was suggested to the subjects, was designed for the controlled production of sensory stimuli.[448] The machinery did nothing at all, but no matter whether they had been led to expect an electric shock, a change in temperature, or the release of an odor, ninety percent of those tested reported the appropriate stimulus when the equipment was set in motion.

This sort of perception without any relevant sensory input is what is usually known as hallucination. It can be most readily invoked in laboratory situations by sensory deprivation in black rooms and immersion tanks, or by chemical manipulation using hallucinogenic substances.[561] But normally an organism does not operate in a sensory vacuum; it is always receiving information of some kind. So the most common form of anomalous experience is one of misapprehension, in which a subject interprets a straightforward situation in an unusual or misleading way, or processes an unusual situation in a more rational, but equally misleading, way.

Donald Broadbent at Oxford tried intentionally to mislead his subjects by feeding through earphones a different message to each of their ears.[76] In one test the left ear might hear the word "red" repeated rapidly three times, while the right one heard "green" repeated at exactly the same volume and speed. If asked to pay attention to the left ear, the subjects reported hearing "red red red, green green green." And when instructed to tune in to the right ear, the same stimulus was heard in reverse order. This led Broadbent to suggest that the nervous system must be considered as a single communication channel of limited capacity; and that it was able to cope with an overwhelming amount of information from a large number of sources, because it possessed a filter which dealt with the bottleneck by processing signals selectively.

It seems likely that the filter is a left-hemisphere system, because it is terribly logical, insisting on acceptable meaning even when none is actually available. If, for instance, the left ear hears "red sails was all sun set" and the right is given simultaneously "green sleeves in the my joy," it is able to make sense of this gibberish. A subject asked to listen with the left will almost certainly pick up "Red sails in the sunset," while a right-ear bias will be rewarded with "Greensleeves was all my joy." The filter juggles the signals and sense around, even switching ears and hemispheres in rapid succession, until they make sense. Reason is imposed on reality and alters it almost at will.

The brain sorts out visual perception in a similar way. In 1962 two American neurophysiologists made the really basic discovery that cells in the visual area of the cortex respond to specific patterns, some to movement in one direction only, some to lines oriented at a certain angle and only that angle, others to corners.[269] The cells function, logically, something like letters which are combined to form neural words, which in turn are arranged into perceptual sentences. But perception is not just a simple passive reflection of external stimuli. What we actually "see" is only a fraction of all the possible visual signals in our environment. The eyes, like cameras, may pick up most of what is there, but even they are selective, because it is the brain that directs their attention and focusing. There is no simple one-to-one photographic relationship between internal perception and external reality. The visual language of the brain is immensely complicated by processes of matching, comparing, analyzing, and a large amount of creative synthesis. The visual letters and words are shuffled around until they agree with certain preconceptions about reality. Someone who puts on inverting eyeglasses will see the world upside down; but if he keeps them on for a while, the brain takes over and reorganizes perception, restoring order so that he once again experiences things "the right way around." Construction and reconstruction carry on to a large degree quite independently of what is actually out there. Perception of all kinds is a function not only of what is coming in to the brain, but what is already there inside.[220]

One can sympathize with the man from Missouri who said, "Show me," and understand why even now it is possible for people to nod their heads sagely and confirm that "seeing is believing." "I wouldn't have believed it if I hadn't seen it" has become a truism; but the structure and function of perceptual mechanisms suggest that it might be more neurophysiologically precise to say, "I wouldn't have seen it at all, if I hadn't already believed it in the first place."

All life is bombarded with an excess of information, and the capacity to tune in selectively on only part of this range has always had survival value. Every species is a sensory specialist, and the more complex ones clearly operate with the assistance of elaborate internal programs. The reality they experience is very largely the one they have selected for themselves. So, given that we are subdivided, rented, and shared, it is not at all surprising that our experience should sometimes be anomalous and contradictory. The marvel is that we enjoy any continuity at all. The apparent consistency of experience is so compelling that it has led us to make an enormous number of unwarranted assumptions about reality. Most fundamental, and most misleading, of these is the basic scientific assumption that there are such things as laws of nature, and that these laws are the same for all observers.

The orthodox rationale goes something like this: The universe is composed of physical matter which is digitized into ultimate particles with distinctive electrical patterns. These units can be considered as data. Transformation of this data takes place entirely through change in the electrical patterns, or, in other words, by natural data processing. And these transformations are under the control of higher programs, written in the alphabet of approximately one hundred chemical elements. This is fine as far as it goes, providing a useful working model for most chemical and physical reactions, but it runs into difficulties with living things.

Life has properties which are independent of the data and have little or nothing to do with the nature of fundamental particles or chemical elements. Factors come into play which may be predicted by the programs and even organized by them, but are ultimately abstract in the sense that they have to do with shape and form rather than substance. The orthodox view encourages mechanistic interpretations such as the cybernetic approach of David Foster, who concludes, "The total universe, inclusive of all aspects of matter and mind, shows a construction virtually indistinguishable from that of an electronic computer."[179] He fortunately has also an essentially humanistic attitude which allows man in this model to foster increased intelligence and happiness by writing his own programs: but the shortcoming of this approach is evident in an insistence that reality is consistent, and that daydreaming is not only delusional, but also dishonest and nonevolutionary.

I prefer to take a softer and more organismic stance and to insist yet again that we have most to learn from the things that don't quite fit, from the events that break the rules and fly in the face of the "laws of nature." Much of the perplexity with which we react to unusual experience is due to logical, camera- or computer-type assumptions about perception and reality. These imply that a percept is a facsimile of what is being perceived, and that memory provides a facsimile of an original experience. Nothing could be further from the truth. We seldom, if ever, function in such an objective, mechanistic way. There is no way one can extrapolate from the laws of nature to "real" events and "true" perceptions about them. There is no single, unquestionable reality and no absolute truth. And this is not just a philosophical conclusion. The evidence of physiology and psychology shows that all our dealings with the world are constructive, interpretive, and tentative.[448] All our knowledge is approximate and relative. At all levels of thought we, and probably all other species that possess any awareness, operate by setting up and testing hypotheses, by solving problems and selecting strategies. And this applies to attention, perception, imagery, and memory, as well as to the more conscious weighing of evidence that we exercise in matters of judgment and belief.

This means that there can be no such thing as objectively right or identical answers. Even the laws of nature are subject to conscious interference, and no two individuals can ever experience anything in the same way. In fact, such is our fundamental disunity and flux that even a single observer can never experience anything twice in exactly the same way. With mind operating on the basis of consensus, constructing, interpreting and juggling carefully selected bits of data around until a satisfactory compromise is arrived at, it is no wonder that our memories are less than photographic and our recall less linear than that of a computer.

The important point in all this is that none of the problems of life and nature have "right" answers. The situation is open-ended, the data are insufficient, the criteria are never absolute, and there are no ultimate solutions. Knowing this, many strange experiences appear much less sinister and inexplicable. And there is no need to be distressed by the fact that we seem to see something which is not there, or to miss something which is.

At the Institute of Psychophysical Research in Oxford, Celia Green has been collecting reports of experiences with apparitions—things that "aren't really there."[216] Her report is totally objective, making no judgments about the events and simply recording what is said to have happened.

Perhaps the most interesting feature of the analysis is that it demonstrates quite clearly that extraordinary events of this kind are usually experienced in a very ordinary way. Two-thirds of the encounters took place in the subjects' own homes, and another substantial proportion in the homes of relatives and friends. Only four percent occurred at work and very few in places that were totally unfamiliar. Ninety-seven percent of all apparitions were experienced without warning or sugges-tion; many of the subjects reported not being aware of any-thing untoward until the person they were seeing or speaking with suddenly disappeared, or walked through a wall. And none of the subjects were aware of being under stress or sub-ject to unusual pressures or feelings; most could think of noth-ing distinctive to say about their emotional state directly be-fore the experience began.

A significant biological feature of these contacts with ap-paritions is that eighty-four percent of the experiences were of a visual nature, thirty-seven percent were auditory, fifteen percent were tactile, and only eight percent were olfactory. The total is greater than one hundred because in some cases more than one sense was involved. The range of the sensory stimuli is wide, showing that all the senses were involved, in-cluding in one case even a sensation of pain; but the involve-ment of each sense is directly proportional to its relative im-portance in our lives.

Eighty percent of apparitional objects in this survey were human beings, most of them familiar to the subjects, and two-thirds of these known positively to be dead. A few were ani-mals, including horses and rabbits, but more often dogs and cats; usually individual pet animals also known to be dead. If there is any substance to the notion that some vital essence survives death, but can nevertheless still be experienced by the living, then it is strange that it should not manifest itself equally everywhere and stranger still that only domestic animals should be capable of survival in this way. And if the apparitions do indeed have some form of independent existence, it is odd that their physical characteristics are differentially realized. All things being equal, we should be able to see, hear, touch, taste, and smell apparitions with equal facility.

The conclusion is inescapable that the vast majority of apparitions fulfill our subjective expectations and cater to our own sensory bias. They are people we know, doing familiar things in predictable places. When dogs are seen, they are invariably reported in faithful association with human figures; cats occur, of course, in a characteristically independent fashion. And almost all the experiences take place when we are comfortable and at ease in familiar surroundings, often lying in bed soon after waking, and therefore most likely to dissociate or to interrupt the conscious stream in some idle and imaginative way. The reports in the Oxford study make it clear that there is a graded series of possible intrusions, ranging from ones in which the apparitional figure merely slides into the normal environment, to other more complex effects in which the subject's surroundings are altered or even totally replaced.

Celia Green calls the latter type of experience "waking dreams," which is probably right. It seems likely that almost all apparitions occur as a result of dream-stream influences during consciousness, or of momentary aberrations in the synthesizer that coordinates and organizes normal conscious perception. Half of all such experiences are over inside fifteen seconds. But even when most apparitions have been accounted for in this way, there is a disturbing residue that are not so easily rationalized.

Most troublesome of these are the ones that appear to more than one person at a time.

Sir Ernest Bennett tells of a landed lady who went one afternoon in 1926 with her steward and masseuse to visit an elderly laborer on her estate.[46] After leaving his sickbed they were returning along the shore of a lake when all three saw "an old man with a long white beard which floated in the wind, crossing to the other side of the lake. He appeared to be moving his arms, as though working a punt but there was no boat and he was just gliding along on the dark water." All three observers saw him, and all agreed that the figure bore a strong resemblance to the old man they had just been to see. Later that evening they learned it was at about this time that he died.

There are many similar ghost stories, but this one is significant because the three people involved gave slightly different accounts of what they saw. The description above is the lady's. The steward saw the old man actually "walking on the surface of the water." And the masseuse saw "a shadowy, bent form step from the rushes, and into a boat." Discrepancies of this kind lead many investigators to infer that one of the witnesses must have been correct, and the others mistaken, or even to imply that all three involved were either misled or dishonest. I, on the contrary, find the differences the most convincing part of the story and am quite prepared to accept that all three actually had separate experiences.

When several people look at the same object, even an indubitably solid one such as a rock, we assume that their perceptions are closely coordinated, like photographs taken from different angles. But we could be wrong. We know so little about how the world appears to any other observer under any circumstances. Colorblindness, for instance, seems to have gone totally unsuspected until the English chemist John Dalton, himself one of the three or four percent unable to distinguish red from green, first described it in 1794. So when three people experience an apparition, it would be astonishing indeed if their reports did not differ in some way.

Hallucinations do occur spontaneously, but the possibility that three witnesses should have similar visions simultaneously, purely by chance, is so minute that it is not even worth considering. Another possible explanation is that one of the three produced the hallucination and communicated it telepathically to the others. If we assume that the various perceptions of the witnesses were no more closely correlated than is strictly implied by their own accounts, then the amount of information which would need to be exchanged between them is actually relatively small. The common elements in their separate experiences could be summarized in a few brief bits of data—"Old man, crossing the lake, that way, now." But a complex hallucination is a rare event, and telepathy, even between two people, seems to be equally uncommon. So the likelihood of telepathic communication of an improbable vision, between three observers, at the precise and coincidental moment that a fourth person died, becomes vanishingly small. And this explanation begins to look a little strained.

We are left, it seems, with only one further possibility: that there was some external cause which provided essentially the same information for the minds of all three witnesses—a genuinely paranormal apparition.

There is, however, a snag with this too. In the literature on apparitions, there are reports of groups numbering from two up to seven or eight individuals seeing something similar at the same time; but there are no well-authenticated cases of larger groups ever doing so.[228] Theaters are notoriously haunted places, and yet no mass audience has ever collectively witnessed an apparition on stage. Why? Well, I am tempted to assume that at least one of those included in the experience of a collective hallucination must be more involved than the others, and is somehow responsible for setting up an external display for their benefit. And that this illusion is in some way energy-dependent and is depleted by the participation of the other witnesses. Perhaps you can't crowd an apparition.

This possibility of tapping into the physical resources of one of the individuals involved could be why so many people who see apparitions feel cold or experience "an icy draft," even if they are not afraid, and despite the fact that others in the room at the time are unaware of any change in environmental temperature.[550] Nobody ever reports feeling hot in the presence of an apparition.

I suggest that groups of individuals act like gatherings of separate cells, and that congregations, under certain conditions and given the right kind of ingredient in the right proportions, reach a critical mass. By virtue of their configuration, their special shape and form, they acquire properties which were not present in the individuals taking part. The force which combines individual people into an organization is as mysterious as that which unites separate cells into a functional whole; but it works in the same way. Small groups may simply interact in an unusual fashion, but complex associations can act like the algal and fungal components of a lichen, and produce something different from either and with quite distinct and novel powers.

The novelist Elias Canetti regards the human crowd as an organism in its own right and has, on this basis, built up a fascinating picture of its natural history.[98] It is conceived by a crowd crystal, which can be an individual, a place, or an occurrence, but is most often, and with most devastating effect, a powerful archetypal idea. The only prerequisite for this seed is that it should be easily recognizable, and capable of being taken in at a single glance. In the right environment, its growth is rapid. "Suddenly everywhere is black with people and more come streaming from all sides as though streets had only one direction. Most of them do not know what has happened and, if questioned, have no answer; but they hurry to be there where most people are." At one moment there may be a few scattered individuals, but introduce a relevant seed, and in the next moment there is concerted action as though the crowd components were joined by invisible bonds. Movement in any part of the organism seems to transmit itself to all other parts, like the waves of nervous discharge in a jellyfish.

The crowd is an organism that feeds on people, and in its juvenile phase it is governed by only one instinct—by the urge to grow, to engulf more people. "It wants to seize everyone within reach; anything shaped like a human being can join it." And, like all children, it tends to be destructive. It enjoys breaking things. "The noise of destruction adds to its satisfaction; the banging of windows and the crashing of glass are the robust sounds of fresh life, the cries of something newborn. It is easy to evoke them and that increases their popularity. Everything shouts together; the din is the applause of objects. There seems to be a special need for this kind of noise at the beginning of events, when the crowd is still small and little or nothing has happened. The noise is a promise of the reinforcements the crowd hopes for, and a happy omen for deeds to come."[98]

In this formation period, the crowd organism is still a delicate thing and can easily dissolve unless its love of noise and density and movement are channeled. As it grows, it has to have a direction, a common goal which strengthens its sense of unity and equality. The nature of the goal determines the species to which the adult crowd will ultimately belong. If it is intent on a killing or an expulsion, it becomes a quick crowd. These are the most conspicuous kind, gathering rapidly and, once they have achieved their goal and reached their point of mutual discharge, which is often accompanied by an almost orgasmic cry, disperse as quickly as they formed.

Less evident, but more influential, are the slow crowds with long-term goals. In contrast to the open, voracious nature of quick crowds, these are closed, renouncing growth and putting their emphasis on permanence. They occupy finite spaces and establish their identity by erecting boundaries, beyond which fall all those who don't really belong. Discharge is delayed, often indefinitely, and in order to keep such a frustrated organism intact it has, as Canetti puts it, to be domesticated. Since it is not getting its usual food, which is more people, a crowd which is not increasing is in a state of fast. And only the great religions and ideologies have been able to develop and master the disciplines necessary to hold out through fasts so long that the goals come to seem almost unattainable. The more idealistic the goal, the more authoritarian the controls have to be to keep such an organism together.

Jung says, "It is a notorious fact that the morality of society as a whole is in inverse ratio to its size . . . Any large company composed of wholly admirable persons has the morality and the intelligence of an unwieldy, stupid and violent animal." His technique of analytical psychology was concerned largely with helping individuals to develop their own personality by coming to recognize not only the content of their own personal unconscious, but also the nature of the forces inherent in collectivity. And he was very well aware of the danger that existed in lifting or abolishing protective repressions too soon. "The forces that burst out of the collective psyche have a confusing and blinding effect. One result . . . is the release of involuntary fantasy, which is apparently nothing else than the specific activity of the collective psyche. This activity throws up contents whose existence one had never suspected before."[294]

Filtered through the mechanisms of dream and consciousness, these contents tend to get distorted into symbols of themselves. Arguably Jung's greatest contribution to our awareness is his careful analysis of patterns which surface in this way, and his organization of them into what he calls "archetypes of the collective unconscious."

This phase of Jung's work began in 1906 in a confrontation with an incurable schizophrenic. He found the patient standing one day at a window, wagging his head and blinking into the sun. When asked what he was doing, the man replied, "Surely you see the sun's penis—when I move my head to and fro, it moves too, and that is where the wind comes from."[302] Though this meant nothing to him, Jung made a note of it and four years later was astonished to come across an identical description in a book on Mithraic religion. He thought it unlikely that his patient could have had any knowledge of a Greek papyrus published years after their conversation, so he assumed that the man had somehow "merged with the mind of mankind" and picked up an original or archetypal image.

There is no theoretical limit to the possible number of such archetypes, but, in practice, analytical psychologists have found that some are more important than others. As usual, the most powerful ones tend to have something to do with sex.

There are, and ideally should be, elements of each sex in the opposite one. "An inherited collective image of woman exists in a man's unconscious," said Jung, "with the help of which he apprehends the nature of woman."[294] He called this the *anima* and suggested that it was a relic of age-old experience and in no way represented the real character of any individual woman. It becomes tangible, he thought, only during the actual contacts with woman that a man makes during the course of his life, and is a potential source of personality problems because it is projected onto those women who attract him. There are naturally discrepancies in the image held by different men, and by men in different cultures, but there seem to be some characteristics that are constant everywhere. The anima has a timeless quality; she looks young, but has years of experience behind her. There is a quality of secret wisdom about her. She is often connected with the earth, or with water, and she may have great power. She is essentially two-sided; one the pure and noble goddess, and the other the seductive witch.

When a man represses his own feminine qualities, it is the dark face of the anima that is most likely to present itself to him, haunting his life with fantasies, moods, presentiments, and emotional outbursts. One doesn't have to look far to find manifestations of this kind in fairies that have the power to entice men away from their homes, in the sirens that lured entranced sailors onto the rocks, in mermaids, water sprites, and nymphs, in malevolent goddesses, *femmes fatales*, and La Belle Dame Sans Merci. The lighter side of the anima carries spiritual values and governs creativity. It is embodied in the muses, in the poetic soul, and in the figure of the Virgin herself. The archetype as a whole is "as thoroughly inconsistent as the woman in whose form she is always personified,"[177] and in describing her in dramatic and mythological terms, Jung conveys the nature of the image more precisely than is possible with any scientific formula.

He does the same for the *animus*, the masculine principle in women, and for archetypes he calls the Hero, the Trickster, the Redeemer, the Dragon, the Old Wise Man, the Great Mother, the Whale, the Monster, and so on. And in each case he shows how a preformed image acts like a magnet, attracting relevant experience to it until it forms a complex strong enough to surface into consciousness. At first an archetype manifests itself only in an individual way. Consider, for example, the God archetype active in a man. As he experiences the world, those things which are relevant to the God archetype become attached to it to form a complex. The complex grows in strength as it accumulates and eventually is strong enough to force its way into consciousness. There it stands a good chance of becoming dominant, so that much of the man's behavior is governed by it. He sees and judges everything in terms of good and evil, he preaches hellfire and damnation for the wicked, he accuses people of sin and demands repentance from them. He believes himself to be God's prophet or even God himself; and eventually others may begin to share his view. He becomes a god and exercises power over his followers that shapes their lives in such a way that a whole community comes to act out elaborate fantasies and experience a reality quite distinct from that which prevailed before this archetype found such extreme expression. A situation of this kind is inherently unstable, and seldom lasts long, but it leaves marks. No community which takes its place will ever be able to ignore it altogether. The archetypes are, it seems, the great decisive forces in history. It is they that bring about events, not our thin skin of personal psychology, our individual reasoning, or our practical intent. Knowing this, and probing daily into the collective unconscious through his patients, Jung in 1918 was already able to warn that "the blond beast is stirring in its sleep and something terrible will happen in Germany."[232]

He was right. This successful prediction and others like it have shown that the model of mind which includes the notion of an unconscious is at least internally consistent and can be used productively to correct specific malfunctions. But nobody has ever seen an unconscious; like the electron, all our evidence for its existence is inferential. We have to take it on faith. This sounds like a mystical proposition, which has no place in an argument that pretends to be scientific; but I would like to stress yet again that we have now arrived at a point in evolution where ideas have acquired a biological reality of their own. Even a purely abstract notion can be acted on by natural selection, demonstrate fitness to survive, and play a concrete role in shaping the destiny of a species.

In the best of all possible worlds, all communication systems would carry the bare minimum of information, because there would be no need to qualify or explain anything. The truth would be self-evident. But things are not like that. Everybody is intent on satisfying the demands of his own selfish genes, and anyone that can get away with it lies. Some beetles, usually known as fireflies or lightning bugs, attract their mates by flashing lights at them. To avoid confusion, each species has its own code, but in North America, females of the species *Photuris pennsylvanica* have discovered that if they lie about their identity by flashing a signal that says "*Photinus pyralis* female here" they can attract and consume the deluded *Photinus pyralis* males.[125] There isn't even honor among thieves. Fledgling birds compete constantly, vying for the attention of their parents, bidding for each arriving meal, loudly proclaiming their hunger even when satiated. There is no such thing as family or species loyalty. Nothing is sacred. Or at least, that is how it seems.

Deception, even within a species, is a biological fact. Whenever the interests of the genes are at stake, there will be lies and deceit; children will deceive their parents, husbands will cheat on wives, and brother will lie to brother. For in the final analysis, it is the individual that matters to the genes. So, very early on in evolution there arose a need for detecting deception in others. Any individual who could avoid being lied to had a distinct advantage.

There was already in the signal behavior of many species a built-in protective device in that the behavior pattern which said, for instance, "I am a receptive female ready for copulation" was so complex that it was unlikely to be initiated with any success by an individual that wasn't physiologically and anatomically in a position to make such a statement. But with the advent of the mammals, behavior has become sufficiently plastic so that male hamadryas baboons *Papio hamadryas* are able to appease other more dominant male troop members by a complex sexual charade that involves adopting the female receptive posture. In many cases the aggressive male who is the subject of this presentation is effectively soothed and even acknowledges the success of the display by a brief attempt to mount and copulate with the deceiver.[587] But close analysis shows that his response is in its own way also highly ritualized, and bears about the same relationship to true mating as a military salute does to the old pattern of actually taking the helmet off a suit of armor and showing one's face in deference to a superior.

Many complex patterns of behavior borrow from other sources. The extravagant courtship display of the peacock *Pavo cristata* can be traced back to feeding movements, so that he is in effect pointing in the most elaborate and exaggerated way at an imaginary food source on the ground in front of him in order to entice the female to him.[475] Shaking, bathing, and drinking movements have been built into the courtship sequence of many species of duck.[396] Territorial defense among African chameleons involves exaggeration of the normal act of breathing into an elaborate display in which the sides of the body are pumped menacingly in and out.[309]

The evolution of almost all sophisticated systems of communication involves extreme opportunism, with signals being borrowed from almost any biological process convenient to the species, and being molded or ritualized into new fixed action patterns. Most of them have become fixed in the behavioral repertoire of the species. In themselves, the movements now have no meaning. A male gray heron *Ardea cinerea* who interrupts his mating to point his head down as though striking at a frog in front of him, and then snaps his mandibles with a loud clash on the imaginary prey, is not necessarily hungry. His hunting behavior has become incorporated as a ritual into the courtship ceremony.[562] It seems likely that its sole function now is to help raise the emotional level of the participants to a point where the actual sexual signals themselves are received with the utmost clarity, and acted on in the most appropriate way.[120]

Human ritual probably serves a similar function, being in itself meaningless to the participants, but helping them to achieve an important goal. Many of our expressive movements, such as smiling and crying, have become ritualized in this way during our evolution and are no longer necessarily directly connected to the stimulus that may once have provoked them. We are all familiar with the child who falls and hurts himself a mile from home, but doesn't actually start crying until he reaches the garden gate. In many cases, the biological base is so distant that it has been totally lost. The stylized movements of dance, the actions of actors in a Japanese drama, the gestures used in the finger language of the deaf, are all now completely symbolic and totally divorced from the emotional state of the individual concerned. They are, in a biological sense, all lies.

Many of these new patterns may ultimately serve the ends of the genes by making their survival machines more efficient, but the important point about them is that they were invented by the phenotype. They are evidence of the increasing emancipation of higher organisms from the dictates of the genetic system. The voice which speaks through them has, in it, distant echoes of old genetic commands, but it throbs even more vibrantly with the new sounds of psychic freedom from the old physical constraints. It is an expression of independence by the contingent system, a mark of the predominance of memes over genes.

A termite hatching from an egg in the brood chamber of a colony, is preprogrammed in every way necessary to allow it to go straight into action as a fully functional member of the community. But a human child emerges into a far more complex environment where things tend to be equivocal, and nothing is necessarily quite what it seems. All it has at its disposal to cope with this baffling situation is an inherited predisposition to behave in certain ways which might help it to learn what it needs to know. It also, fortunately, has a long period of total dependence. During this time, an infant's wide-open responsiveness becomes narrowed down by interaction with its mother. Communication between them depends in part on instinctive social releases, but mostly on less stereotyped, more subtle and individually variable signals. Gradually, a baby learns new ways of extending its range of communication, only a small part of which is directly connected with physiological needs. Both mother and child seem to enjoy communication for its own sake, and the behavior of each is clearly regulated by the behavior of the other.[74] But the infant has to take a large amount on trust. There is evidence that there is in it an early disposition to comply with maternal commands and prohibitions, and that this is independent of training or discipline.[506] This makes it possible for a baby to learn that she, on whom it depends utterly, is reliable, that a mother temporarily absent, and therefore nonexistent, is not necessarily lost forever. This now-you-see-her, now-you-don't experience with mothers is almost certainly our first contact with the numinous, with something inexplicable and awe-inspiring. It teaches us to trust long before we learn to speak. It demands of us our first demonstration of faith in something which cannot be known, whose existence cannot be proved by normal means. It is our first contact with the world of abstract ideas; our first experience of the sacred.

Some of the information that comes into our lives presents no problems. It is either obviously true, or it can be assumed on the basis of our experience to be true. But most of the messages on which social actions rely are neither logically necessary, nor can they be validated from experience. We have to take them on trust. Initially it was possible to believe something simply because "Mother said so," but this rationale becomes a little flimsy for anyone over the age of ten. So there obviously arose in human evolution a need for some other kind of validation for many mysteries. Roy Rappaport of the University of Michigan suggests that sanctification supplied this need. He says, "People are more willing to accept sanctified than unsanctified messages as true; to the extent that they do, their responses to sanctified messages will tend to be predictable and the operation of the society orderly. The acceptance of messages as true, whether they are true or not, contributes to orderliness and may, in fact, make it possible."[447]

The abstract concept of sacredness precedes language in any individual, and could even have been responsible for the development of all symbolic communication and the rapid development of our intelligence. And it seems clear, however you look at it, that it was intelligence and not genetics which was responsible for our elaborate social organization. The conventions which hold society together are, however, largely arbitrary, and the very intelligence which made us as a species adaptive enough to learn an extraordinary variety of widely differing social patterns also makes it inevitable that we will question this arbitrariness and look for alternatives. But no society, if it is to avoid chaos, can allow all alternatives to be practiced. So, in addition to genetically determined individual selfishness and deceit, human societies are faced with what the vitalist French philosopher Henri Bergson called "the dissolving power of intelligence."[50] It seems to have been primarily to fill this gap that religion was born.

The virtue of regulation through religious ritual is that the activities of large numbers of people can be governed without powerful human intervention, or indeed without any human authority of any sort. Sanctity is much less expensive, and far less divisive, than a police force. By setting up something as sacred, the needs of society are presented to the individual as though they were his own goals, and possible disruption is easily and quietly replaced by unquestioning compliance. This makes it sound as though there must be some godlike figure manipulating evolution towards a distant but divine goal. There may indeed be, but it is not necessary to make such an assumption. The programs which acquire sanctity and do come to be followed can have purely ecological origins.

In the Highlands of New Guinea, pigs are an important part of subsistence economies. They provide protein, convert garbage, dispose of feces, soften the soil for gardening, and function as wealth objects used in bride payments and in compensation for headhunting raids. They are crucial to the societies, but there are times when they can become too much of a good thing, invading gardens, taking up unreasonably large amounts of time and effort, and destroying the environment. So, over thousands of years, an elaborate pattern of sacred ritual has arisen around pig husbandry. Whenever the pig population reaches a certain critical level, an elaborate year-long ritual called the *kaiko* is set in action. The centerpiece of this cycle is the sacrifice of large numbers of pigs to the "red spirits" of the ancestors, and ostensibly it is done entirely for their benefit. But the net result of each *kaiko* is that the donors gain prestige and some relief from the tribulations of pig-keeping, the recipients acquire high-protein foods, the frequency of warfare is regulated, people and land are partially redistributed, and the environment is protected from despoliation by a plethora of pigs.[446]

The existence of rituals such as this can be explained in biological terms. Over long periods of time, natural selection will inevitably favor those ceremonies and beliefs that have the greatest survival value. All others will be selected against by default. Ideas are subject to the same evolutionary process as beaks or claws. They have to prove themselves in action. Memes compete for survival in exactly the same way as genes, and only the fittest survive. But it is important to appreciate just whose survival we are talking about. An idea can be inimical to society, to individuals and to the genes themselves, and still survive. It could even, in the form of mechanical memory in a book or film or computer record, survive the total extinction of our species. It evolves and comes to be what it is, simply because the characteristics it has are advantageous to itself.

I suggest that Jung's archetypes should be viewed in this light. Many of them obviously have social and even environmental value; they act as psychic checks and balances, shaping us, controlling individual personality and communal concerns. These probably have ecological origins and ought to be susceptible to analysis in the same way as the *kaiko* custom in New Guinea. I predict that a new breed of biological anthropologists, skilled also in analytic psychology, will soon take the field and begin to document the functional role of such archetypes in our lives.

There are other archetypes whose existence cannot be so easily explored. They lurk in the area Jung identifies as the collective unconscious, and I suspect that his intuitive feelings about them and it were probably close to the truth. It is possible to interpret some patterns of normal human behavior and some psychological malfunction with reference to them, and it may even one day be possible to produce evidence of their independent reality, but at the moment they remain in an area which is still not in open discourse and can neither be proved by experiment, nor discredited by reason.

There is, however, one phenomenon which seems to be an integral part of this mystery zone and on which sufficient evidence has already been collected to make at least a preliminary analysis. I mean the persistent and apparently timeless belief that we have been, and still continue to be, visited from the air by entities from legendary countries, separate realities, or other worlds.

The literature is vast and mostly partisan. I find much of it difficult to read, but it has become impossible to ignore. Jung felt he had to deal with it somehow and, as usual, put his finger on a sensitive spot. In 1954 he wrote an article in a Swiss weekly in which he expressed himself in a skeptical way, "though I spoke with due respect of the serious opinion of a relatively large number of air specialists who believe in the reality of UFOs." When this article was discovered by the world press, the news spread rapidly that Jung was a saucer-believer. He immediately issued a statement giving a true version of his opinion, but this was totally ignored. So he published a fuller account in 1958 with a preface in which he noted, "As the behaviour of the press is a sort of Gallup poll of world opinion, one must draw the conclusion that news affirming the existence of UFOs is welcome, but that scepticism seems to be undesirable . . . there is a tendency all over the world to believe in saucers and to want them to be real."[300] There is, and quite apart from the question of their reality, it would be fascinating to know why.

From the morass of ufology, one voice rises clear and clean —that of Jacques Vallée, a French astrophysicist. He defines the area as that concerned with "the myth of contact between mankind and an intelligent race endowed with apparently supernatural powers."[555] And he points out that our present perplexity is by no means unique. "However strong the current belief in saucers from space, it cannot be stronger than the Celtic faith in the elves and the fairies, or the medieval belief in lutins, or the fear throughout the Christian lands, in the first centuries of our era, of demons and satyrs and fauns." There certainly seems to be a strong similarity between the entities described as the pilots of UFOs and the fairy folk, elves, and sylphs of the Middle Ages. Vallée points to the example of the *fadets*, hairy little black men who were said to live in caves in the Poitou region of France, and who, as recently as the mid-nineteenth century continued to play tricks on terrified women. That same area today is free of *fadets*, but it is one of the hotspots in Europe for UFO sightings. Similarly, in northern Mexico there are said to be one-meter-tall, hairy, black humanoids, known locally as *ikals*. For centuries these "spirits of the air" have been attacking people on country roads, but today they do so with the aid of rockets strapped to their backs.

I mean no disrespect to either UFOs or fairies by equating them in this way, but must agree with Vallée when he says, "Attempting to understand the meaning, the purpose of flying saucers . . . is just as futile as was the pursuit of fairies, if one makes the mistake of confusing appearance and reality." Which does not mean that those who experienced either were necessarily suffering from hysterical delusions, but it is interesting that to observers in the United States, the visitors tend to appear as mechanical contraptions or as science-fiction-movie monsters; while in South America, they are typically bloody-minded and quick to get into a fight; whereas in France, "they behave like rational, Cartesian, peace-loving tourists."[555]

There is no escaping the conclusion that there is, at the very least, a strong psychic component in the phenomenon. And there is no avoiding the parallel observation that the majority of UFO enthusiasts tend to be committed to a physical explanation, and sometimes go to extraordinary lenghts to belittle or suppress reports that make the experience of an object appear too fantastic or bizarre. It is easy to understand their fear that outlandish eyewitness accounts could be prejudicial to any chance they might have of putting their study on a sound scientific basis. Such concern is undoubtedly well founded, but it is well to keep this nuts-and-bolts bias in mind when analyzing reports, because it tends to conceal the more whimsical, archetypal nature of much of the UFO experience.

And yet, there is every reason to be equally fascinated by the apparent physical reality of many appearances. I have already suggested that most, if not all, apparitions and ghosts could be tracked down somewhere within the nervous system, but this is not true of UFOs. Some can be picked up on radar, some leave traces on the emulsion of photographic film, and one seems even to have made two passes at the Ethiopian village of Saladare in 1970, knocking down houses, tearing up trees, fusing tin roofs into unrecognizable blobs of metal, and melting masses of asphalt on a nearby road.[522] Reports of actual landings are often supported by scorched, sometimes radioactive, earth, by circular flattened areas in grassland or cultivated crops, and by deep indentations in the ground. On at least one occasion in New Mexico, metal scrapings from a desert landing site were analyzed and shown to be of an unusual alloy.[503]

This is no place for an exhaustive review of UFO cases, but a brief summary of the facts is instructive. There is very little uniformity. UFOs are shaped like discs, spheres, cylinders, cigars, dumbbells, ovals, eggs, diamonds, cones, parachutes, tops, mushrooms, and hamburgers; all of these shapes occur singly or in groups. They range in size from a few inches to more than a mile in length, and travel fast or slow, with or without undulations, wobbles, zigzags, or sudden changes of direction. They are smooth, hairy, knobbly, shiny, dull, every color of the rainbow, and come equipped with optional landing gear, wheels, tripods, doors, and windows. They are completely silent or else they hum, whoosh, hiss, flutter, whine, whistle, beep, pulse, buzz, vibrate, bang, blast, roar, or explode thunderously. They stop internal combustion engines, interfere with radio and television transmissions, shock, burn, and paralyze people, and cause them to lose consciousness; or not as the case may be. In short, they are all things to all people.[391]

Since 1946 there seems to have been an acceleration in both interest and in manifestations. Research groups have sprung up all over the world to collect local data, and these have now been pooled and analyzed in a variety of ways.[554] Aimé Michel in France finds alignments in the sightings that produce straight line, star-shaped, and even geodetic grand circle designs.[374] Bruce Cathie in New Zealand links them into a mathematical grid whose coordinates are based on the harmonic relationship of gravity, the mass of earth, and the speed of light.[102] Jacques and Janine Vallée have looked for patterns in time and found cyclic variations that seem to correspond with planetary movements and terrestrial seasons.[557] Ivan Sanderson, like a good biologist, finds significance in the behavior of the objects seen.[474] Allen Hynek in the United States concentrates on observational categories and close encounters experienced by "at least two persons of demonstrated mental competence."[275] Gordon Creighton in Britain is more concerned with direct contacts and the character of the alleged aliens themselves.[118] All these surveys, no matter what their approach or concern, seem to have enjoyed precisely the same kind of limited success. There are patterns to be found in the geography, history, meteorology, ethnology, age, sex, and anatomy of the objects and the subjects of the UFO experience; and these cannot be accounted for by chance alone. All the findings are suggestive, but none are ever conclusive.

As a biologist myself, I am particularly interested to find that we are not the only species involved in the experience. Dogs, it seems, are the best UFO detectors, and dislike them intensely. Some of them bark, some howl, some froth, and some cower in terror when an object is about. And all these reactions are elicited before humans are aware of anything unusual. It may be a high-pitched sound which alerts them, or it may be microwave transmissions, but the chances are that the phenomenon is as variable for them as it is for us. Many other species are equally unenthusiastic. Cats hiss and spit, sheep stampede, horses rear up, cows lie down, and birds simply stop singing. Most reactions seem to be temporary, but a few persist. Cattle refuse for several days to be herded into paddocks over which UFOs have been seen to hover; and in one case a single sniff at a recent landing site sent a dog dashing away howling.[15]

At the time Jung wrote his paper, he did not have access to this wealth of quantitative material. He concluded nevertheless that "we are dealing with an ostensibly physical phenomenon distinguished on the one hand by its frequent appearances, and on the other by its strange, unknown, and indeed contradictory nature." And he decided that the evidence could support one of two hypotheses. "In the first case an objectively real, physical process forms the basis for an accompanying myth; in the second case an archetype creates the corresponding vision."[300] Nobody will be surprised to learn that he came down in favor of the latter.

Jung called the UFO reports "variations on a visionary rumor," and likened them to the collective visions experienced by crusaders during the siege of Jerusalem, by troops at Mons in the First World War, and by the faithful gathered to see the Virgin at Fatima in Portugal. And, he reasoned, "if it is a case of psychological projection, there must be a psychic cause for it. One can hardly suppose that anything of such worldwide incidence as the UFO legend is purely fortuitous and of no importance whatever." He was particularly impressed by the fact that so many witnesses were not expected to see anything, had no previous belief in UFOs, and were in general characters renowned for their cool judgment and critical reason. "In just these cases," he observed triumphantly, "the unconscious has to resort to particularly drastic measures in order to make its contents perceived. It does this most vividly by projection, by extrapolating its contents into an object, which then reflects back what had previously lain hidden in the unconscious."[300]

He went on to suggest that such individuals were involved in a collective experience because "a political, social, philosophical, and religious conflict of unprecedented proportion has split the consciousness of our age." Jung believed that tension of this kind created a potential which often expressed itself in a manifestation of energy. He had sound personal reasons for holding such a belief because in 1909 in an altercation with Freud, a loud detonation had occurred in a bookcase at their side; closely followed, when Jung, who felt a great inner certainty about it, said, "I now predict that in a moment there will be another loud report," by a second similar explosion.[301]

And because most of the UFOs known to him were saucers, discs, cylinders, or spheres, Jung interpreted them as symbols of unity and wholeness, rising spontaneously out of the unconscious to provide a focus for healing.

Jung wrote this just three years before his death in 1961, and since then the global situation has gone on deteriorating and UFO manifestations have become even more varied and bizarre. There is new impressive evidence of physical reality, but there has also been a growing awareness of the paranormal potential of purely mental energy. So the most interesting and productive of recent analyses all seem to be following, and adding to, Jung's conjecture that the phenomenon has a psychic origin. Ronald Grunloh, an anthropologist, proposes that "instead of ascribing Ezekiel's experience with the fiery cloud, Moses' conversation with the burning bush, and Lot's confrontation with the angels, to the landing of flying saucers; we try instead to explain the sighting of flying saucers as experiences of a kind similar to the religious visions of the past."[227] The ever-reliable Jacques Vallée suggests we are dealing with a control system, something like a thermostat, which sends up cycles or waves of UFO experiences to reinforce certain conscious attitudes and generate rapid psychic change. He hasn't yet decided whether this program of classical conditioning is designed by someone else, or is self-imposed.[556] David Tansley, an English chiropractor, sees UFOs as signs of mind interacting with other natural forces, "enticing man to look at his world in different ways, triggering off new ideas and states of consciousness, and filling man with feelings of awe and wonderment each time they observe these omens of awareness."[522] It looks as though ufology is about to become a much more spiritual pursuit.

I'm sure that's the right approach and I hope it will be productive, but I am uneasy about some growing trends. It was probably inevitable that the inability of even the most dedicated and objective researchers to provide final, unequivocal proof of the physical existence of UFOs would engender disillusion; and logical that this failure should produce a pendulum swing to more psychic solutions. But there has been the usual mindless rush for easy answers, and two rival schools have come into existence, each latching onto one of the two oldest scapegoats in the mystery business, God and the Devil.

Erich Von Däniken is more or less responsible for the first approach with his attempt to explain many of the marvels of antiquity as the work of extraterrestrial visitors.[564] I find his thesis bleak, and insulting to our ancestors, whose accomplishments he constantly demeans; but I have to admit that he produced an extraordinary response, and sparked off an interesting and challenging discussion. That's all to the good, but unfortunately he asked the provocative question "Was God an astronaut?" and his aliens, and others in a string of books that followed, came to be referred to as gods. Flying-saucer cults sprang up like magic mushrooms all over the globe, many of them explicitly religious in their structure, often dominated by one person who claimed to be in direct contact with the source of celestial intelligence. During 1976 a missionary couple called Bo and Peep began preaching deliverance in the United States and even persuaded hundreds of their followers to sell everything, leave jobs and families, and gather at assembly points for transport to other-planetary paradises. They were disappointed. This kind of development is reminiscent of the millenarian cults which sprang up in post-feudal Europe as agricultural economies began to break down, and mobs of displaced peasants gathered in expectation that the heavens would open and receive them. They too were disappointed, and many more will face financial and emotional ruin if this trend persists.

The polar position seems to have arisen in direct response to the charioteers, but it offers little solace. The approach is exemplified by Hal Lindsey, Thomas McCall, and John Weldon, self-confessed Christians and biblical scholars, who see all around them dire prophecy come to pass and signs of the end of times.[349] Like Von Däniken they have voracious and selective appetites for the facts that happen to fit their theory. The restoration of Israel and the prevalence of earthquakes are proof that "the stage is set for earth's final act . . . watch for the Antichrist . . . his appearance will mark the beginning of your final seven years."[390] The current interest in the occult, research into parapsychology, and above all the abundance of UFOs show that "demons are responsible . . . UFO entities are evil, hostile beings . . . and the worst is yet to come."[576]

Prophetic and millennial cults on their own seldom survive long, usually disintegrating when the deadline passes uneventfully. A true religion needs, in addition, to have some kind of enduring social utility. Visions, whether they be of earth spirits, flying saucers, or the Blessed Virgin Mary, are not enough. The shaman who communes with earth brings back power to lead his people through crisis. The Christian saints survive by virtue of the constructive impact of their work on society. So what is keeping the fairy-phantom-UFO complex going for so long?

As a meme, the idea of visitation from other worlds has survival value. It persists because it meets a deep-felt need. Perhaps because it tells us the truth about ourselves?

I don't mean that beings in spacecraft landed here and created man in their image; at least not literally. The Von Däniken thesis, as it stands, can and has been attacked on grounds of sloppy thinking.[516] It is filled with factual and logical errors, and persistently suppresses the abundant archaeological evidence that even the examples he chooses were the unaided work of humans like ourselves. But I don't believe that the popularity of his books is due simply to the fact that millions of readers have been waiting impatiently for someone to come along and shatter the conventional theories of history and archaeology. There is a certain satisfaction in any situation where an amateur succeeds in confounding the experts on their own ground. Superficially, Von Däniken seems to have accomplished that, but I believe there are more fundamental reasons for the fervor of his following.

If you multiply the Von Däniken time scale by one thousand, if you substitute the mystery and intricacy of living things for the marvels of ancient architecture, if you can stretch your conception of spacecraft to include meteors, comets, and conglomerates of interstellar dust—then earth was indeed visited, and continues to be visited, by extraterrestrials who have been largely responsible for the existence of mankind.

And I suggest that this knowledge is deeply embedded in the biological unconscious, and is one of the things with which it is most lastingly concerned. If it becomes possible to prove Fred Hoyle's notion that earth is intermittently invaded by disease-producing organisms, we might find that the periodic arrival of quantities of these is directly connected with outbreaks of UFO activity. It might even now be worth looking for correlations between saucers and sneezes; but I suspect that the collective unconscious doesn't need much prompting, and that it can send up evidence of its concerns in response to a variety of stimuli, or even in the absence of any trigger at all.

If there is anything in this idea, then why should this be all that biological memory has on its mind? It isn't all, by any means. We have I think, in the growing catalogue of Fortean phenomena of all kinds, evidence of many similar concerns.[376]

Ancestors in the air? Why, certainly, sir. And on September 23, 1973, tens of thousands of tiny toads showered down onto the southern French village of Brignoles.[17] Frogs, fishes, rats, snakes, jellyfish, blood, and flakes of meat have been falling, from thunderclouds and clear skies alike, at random intervals for as long as there has been anyone to keep records.[178]

Fiery reentry into earth's atmosphere? Nasty. On December 13, 1959, Billy Thomas Peterson, aged twenty-seven, of Pontiac, Michigan, while sitting in his car, is consumed by a fire that turns most of his body to ash, but doesn't even singe his underwear.[12] Spontaneous or preternatural combustion of human beings is something that happens often enough for it to have been given a medical description—"auto-oxidation." The author Michael Harrison, who has collected records of hundreds of cases, describes it as starting with "a bluish flame being seen to extend itself, little by little, with an extreme rapidity, over all parts of the body affected. This always persists until the parts are blackened, and generally until they are burned to a cinder. Many times attempts have been made to extinguish the flame with water, but without success . . ."[239]

Enemies in the wind? Certainly. On several occasions in August 1960, a South African police chief and three of his constables watched helplessly while twenty-year-old Jimmy de Bruin was cut deeply on the legs and chest by an invisible assailant with an equally insubstantial scalpel.[148] Since 1973 mysterous mutilations of cattle, sheep, and horses, many of which are found totally drained of blood, have been taking place in Colorado, Utah, and Montana. One researcher puts the number of cases now at something like seven thousand.[41]

Antique struggles with other species? Everywhere. On August 15, 1966, a twenty-five-foot-tall ape terrified the residents of the little Malaysian village of Segamat.[13] The world abounds in monsters, giants, sea serpents, demon dogs, phantom cats, mothmen, and other motley creatures from the black lagoons of the mind.[313] The psychologist Stan Gooch is probably right in his contention that modern man is a hybrid drawn from both Cro-Magnon and other sources; and that many of our bad dreams, the visions of trolls and hunchbacked dwarfs, are memories in the modern part of our mind of our Neanderthal origins.

Part of us seems determined to frighten the rest of us to death. And yet, if the popularity of horror films, ghost stories, and ghoulish goings-on is taken into account, we not only put up with these hauntings, but actively seek them out. John

Napier of London University suggests that "monster worship has both a learned and an inherited component: the potential is inborn, but the object of the worship has to be discovered by learning . . . Man admires bigness for its own sake, and if physical bigness is linked with psychological bigness, whether of natural or supernatural origin, then the combination cannot fail."[400]

I have absolutely no hesitation in assigning much that now amazes and disturbs us to the common cause of a biological unconscious. From this base flow beliefs and superstitions, customs and myths, folklore and fallacy; and all are protean entities constantly changing shape as they adapt to current needs and the contemporary scene. "Flying saucers may be a modern phenomenon, but they come from the same stable as Pegasus, the flying horse of Mount Helicon."[400] They feed the same need.

And yet it is not enough to fall back on basic biological dreams and cultural clichés. We know our psychic monsters also by their physical productions. UFOs scorch the earth; the Loch Ness phenomenon is communal and photographable; the yeti and the Sasquatch leave footprints in the snow. The one constant feature that all seem to share is their reality status, which is more than purely psychological and something less than hard. There is never quite enough evidence to satisfy critical scientists and yet often just enough to keep some of them coming back for more. This elusiveness of all the relevant phenomena may be the best clue we have to their true nature.

It seems that we have made a dreadful mistake by identifying ourselves exclusively with our consciousness, by imagining that we are only what we know about ourselves. There is much more to us, parts that speak in tongues that take some getting used to. To learn their language, we need to lend a very attentive ear to the strange myths of the mind, and take a careful look at whatever happens, regardless of whether or not it accords with orthodox explanations. For, as Jung says, "it is important and salutary to speak also of incomprehensible things."[301]

# PART FOUR

# A Myth of Our Own

Man in an invisible act of creation put the stamp of perfection on the world by giving it objective existence. This act we usually ascribe to the Creator alone, without considering that in so doing we view life as a machine calculated down to the last detail, which, along with human psyche, runs on senselessly, obeying foreknown and predetermined rules. In such a cheerless clockwork factory there is no drama of man, world, and God; there is no "new day" leading to "new shores," but only the dreariness of calculated processes.

What we need is a myth of our own.

—CARL GUSTAV JUNG,
*Memories, Dreams, Reflections*

In 1799 a boy was discovered running wild among the rugged plateaus of Rouergue in southern France. Known, after his capture, as Victor of Aveyron, he came under the care and tuition of the great educationalist Jean Itard. But despite years of effort, the professor was never able to coax his charge into an acceptance of prevailing custom. He showed so little response to all approaches that for a long time it was assumed that he was deaf, until one day, on a walk through the forest, he was seen listening avidly to the small sound of a mouse rustling in the leaves.[280]

We all tend to respond only to what we recognize. What we choose to recognize—because, in the final analysis, we don't really see with our eyes or hear with our ears. Perception is an aspect of behavior, and behavior is organized in the brain. And though all human brains are superficially similar, different people use different parts of their joint vast potential. Our lives and our understanding of the world, perhaps even reality itself, are influenced by ecological factors.

The human eye has what is known as an anisotropic acuity pattern. Early research established that it was generally better at discriminating horizontal and vertical lines than ones that were slanted; and it was assumed that this was an innate, genetic characteristic of the human nervous system. But a recent study with Cree Indians from James Bay in Canada has shown that there are cultural differences in acuity.[10] The Cree can discriminate slanted lines as accurately as any others, and the reason for this capacity seems to lie in the nature of their early experience. They live during the summer in a cook tent or *meechwop* and in the winter in a more substantial lodge or *matoocam*. Both are built on wigwam lines with a skin cover supported by an elaborate structure of poles erected at every conceivable angle. So a Cree infant tunes its senses on an environment rich in contours, while most Westerners live and grow in more "carpentered" surroundings with a preponderance of vertical and horizontal forms.

These are sound ecological adaptations with good survival value, both for the Cree child learning to hunt in the amorphous *taiga* forest, and for the ghetto kid struggling to stay alive within the rigid geometry of a city block. We learn to fit the world as we find it, but our nature is shaped not only by physical factors. We are also at the mercy of our cultures, and these tend to bend reality until the world comes to look the way we have learned to talk about it.

It is estimated that the human eye can distinguish as many as seven million different shades of color, but in practice our actual sensitivity tends to be limited to the ones we can put a name to. People in different cultures even seem to divide the visible spectrum up differently, actually seeing a rainbow each in their own peculiar way. It seems to be only since Isaac Newton's experiments with prisms that we in the West enjoy red, orange, yellow, green, blue, and violet in the play of sunlight on drops of water; and only the purists, usually those with a formal education in physics, insist on adding indigo to this list. Aboriginal rainbows, on the other hand, have but three or four colors in them. It isn't easy to experience the unnamed, the unfamiliar. And the things we have named tend to take on the characteristics we expect of them.

The Northern Ojibwa Indians of Michigan classify several quite different wild fruits as *kinebikominin* or "snake berries," believing them to be poisonous.[235] All are in fact perfectly edible, but in these people's minds they have come to be comparable objects of taboo and are even seen to be physically alike, though they have no similiarities that would catch the eye of a botanist. On the other hand, where all we see is snow, Eskimos are aware of at least seven different kinds, each with its own distinct properties and behavior.

We have come to deal in symbols, words and pictures, which structure reality by helping to give things identity, clarity, and definition. A name is much more than just a label, a decal which can be applied or removed at will. It contains meanings and associations, values which transcend the simple frequency and amplitude of the sounds involved. The sounds of speech are nothing like an alphabet anyway, but have their own unique dynamic form which gives them properties no other sounds possess. In listening to speech, we can take in as many as thirty phonemes per second, which is far more than the temporal resolving power of the ear actually allows. Every time we listen to someone speak, we break the rules, we accomplish the impossible. The mechanism is still a mystery, but it seems that the particular phonemes used in speech—and these are the same small selection in all known languages—are capable of carrying information not only about themselves, but also about the acoustic cues which precede and follow them. This is why we need a special speech center in our brains, a decoder which compensates for variation produced by the context in which the speech occurs, glosses over the fragmentation produced by our vocal apparatus, and tracks down and rescues the relatively indistinct parts of the signal which carry the actual message.[346]

All our experience is governed in this way. Our senses mold and shape information to suit their own ends, and our brains shuffle the bits around until they fall into some acceptable pattern. For organisms as complex as ourselves, there is no longer any possibility of pure awareness. Our grasp of reality, our consciousness, is awareness that has been processed through the mill of our needs and our beliefs. In a very real sense, what we regard as ordinary consciousness is a state of grand illusion. Nothing is quite what it seems.[425]

Although it is difficult to say what a baby's mind is doing, it seems likely that it starts with no very clear distinction between itself and the external world. At first there is nothing but experience. As Joseph Pearce puts it, "The child's mind is autistic, a rich texture of free synthesis, hallucinatory and unlimited. His mind can skip over syllogisms with ease, in a nonlogical, dream-sequence kind of 'knight's move' continuum."[428] He nevertheless is programmed by his genes to pick up things as fast as possible, and the fact that we succeed quite quickly in molding him to respond to our criteria is a credit to the flexibility of mind. It doesn't prove that our constructs are the right ones. They simply happen to be the only ones on offer, and all too soon he reaches an adjustment; he accepts our consensus about reality and has to abandon the knight's way; he succumbs to the pressures that make pawns of us all.

So a child moves from direct experience to a knowledge of the world based on a growing series of internal maps. There's nothing wrong with maps as such—they are useful aids to navigation—but the danger is that they tend to work too well. In the end we forget that they are no more than convenient abstractions, and we begin to regard them as the truth. We come to believe in them so profoundly that if something happens which cannot be located on the map, we dismiss it as an error of perception, or a conjuror's trick. We talk ourselves out of direct experience and into an attitude which regards experience as a minor side effect of the laws of nature, which we consider to be totally independent of us.

They're not. The new physics is quite clear on this matter.

There is no longer room in the most advanced maps for a detached objective observer. John Wheeler, one of the pioneers of modern quantum theory, even suggests that it is now necessary "to cross out that old word 'observer' and to put in its place the new word 'participator.' In some strange sense the universe is a participatory universe."[581] We are all involved.

So it should now come as no surprise to us that there can be periodic manifestations of mind popping out all over our reality. That every now and then, in a swirl of the dark waters of the tide, a symbol should float to the surface, embellishing our tidy designs with one of those archetypal monsters from the margins of an old map.

The real mystery is why it doesn't happen more often, why the phenomena should still be so rare and so difficult to control, why we should still be subject to the prohibition on knowing who we really are.

In this final section, I want to explore the possibility of provoking the biological unconscious, meeting it on its own ground and examining what it actually says and does. I want, in effect, to set myself up as God's Analyst and explore the possibility, and the advisability, of releasing the repressed forces of the Lifetide and turning them to our conscious advantage.

# CHAPTER TEN

# EVOCATION: Summoning the Past

The Croatian-American inventor Nikola Tesla, who was, among other things, responsible for designing and perfecting equipment that made alternating current practical, had superb control of his mental imagery. He did all his designing in his head, without working drawings of any kind, but was nevertheless able to instruct a dozen different machinists how to make each separate part, to the nearest thousandth of an inch, so that all the components fitted together perfectly. He was even able to test his machines without actually building them, by putting the imaginary parts together in his head and letting them run there. Thousands of hours later, he would resurrect the image in his mind, stop the machine, dismantle it, and inspect the parts for wear to determine whether they needed reinforcement or redesign.[409]

We rightly recognize such talent as unusual and set Tesla aside as something special, a genius. He was in fact unusual in so many ways that some classified him as "hopelessly neurotic," but labels of this kind are not very useful. They disguise the fact that we implicitly project the working patterns of our own minds as a standard for all minds, and assume without justification that we "normal" people all share a common state of consciousness. We speak a common language, which stresses external rather than internal events, and this seems to give us access to the same experience, but it actually conceals enormous individual differences. And ignorance of these has made it possible for us to ignore the fact that the common ground we do have exists at an altogether different level.

All of us have exceptional talents, but it sometimes takes special techniques to reveal them. Some years ago I was involved in a study on perception which used a modified version of the old game that involved looking at twenty or thirty objects on a tray for a minute, and then writing down as many as one could remember. For our updated version, we devised a whole room filled with a vast collection of unrelated objects and information, including projected film and a series of changing sound signals. Subjects each had a minute's exposure to this multimedia show and, following a five-minute pause, were given written and oral tests of memory. As expected, there were vast differences in score, and some correlation of these with different personality types. All of the subjects were then hypnotized and asked in this state to try again to describe the experimental room. To our surprise, the difference in scores was almost completely ironed out. All but one of the subjects demonstrated an equal and very much greater ability to remember just about every item in the test. All of us, it seems, even those who claim to have poor memories, do in fact perceive and retain vast amounts of information of which we have no conscious knowledge.

The exception in this test was a male student nineteen years old, who produced evidence of a truly extraordinary talent which we might have missed altogether if I hadn't quite by chance asked him a stupid question. On the far wall of the room, twelve feet from where the subjects stood, was a framed copy of the front page of a famous newspaper taken from a time when it used still to carry its personal columns on the cover. This student hadn't consciously recalled it, but under hypnosis he included it as "Paper in a frame." I pressed for more information by asking, "What was in it?"—hoping he might mention that it was a newspaper which, if he had reasonable eyesight or knew the house style, he could identify as *The Times* of London. Several other subjects had done so, and one with excellent eyes had even been able to see, and remember, the headline printed in its upper left corner. But this subject shook me to the core by beginning to recite, just as if he were actually reading it at that moment, masses of detail from columns of fine newsprint that it is humanly impossible to resolve from a distance of twelve feet. Try it and see. Even a vulture, with its special telescopic acuity, probably couldn't manage it.

Some people, most of them children, have what is known as eidetic imagery. They are able to retain persistent images of visual scenes which they "see" as being somewhere "outside the head" and are able to describe in detail by moving their eyes to and fro, just as if they were scanning the original scene.[25] In one study, such a subject was presented with a pattern of apparently random colored dots to look at with the left eye; on the following day a similar pattern was shown only to the right eye. Both pictures were very carefully designed so that when one was superimposed on the other, and only then, a square appeared in the center, apparently standing out from the background. To appreciate this effect, the subject had to conjure up an accurate mental image from the previous day and then superimpose it on the new pattern stereoscopically so that the two fused to form the square design.[518] What my student accomplished was something similar, but he was able also to employ some kind of mental telescopic lens with a variable focal length, and zoom in on the image in his mind in order to read the fine print.

The implications of studies like these are clear. Our personal unconscious, whatever it might be, contains an embarrassing amount of information. And there can be no question of lumping all its contents together as repressed material which is too sensitive or too traumatic to be brought to conscious attention. No one can be that neurotic. Freud's early notion of a censor protecting us from ourselves in this way must now be considerably revised. Perhaps it would be best to refer to the mechanism simply as a filter, and to consider its action as essentially similar to that of the sense organs, each of which allows us only a limited window on the world, structuring our experience and keeping reality within bearable bounds.

There must have been a time in the course of evolution when this kind of protection was essential for an organism new to awareness, with a delicate identity to defend. But it seems we no longer need such rigorous security measures. The barrier is becoming semipermeable, and allowing more and more information through. I believe that only man has reached this stage, and that but recently. I suggest that talents like telepathy are not only new to us, but new to the whole living system. They are not vestiges, as Freud suggested, of a once powerful nonhuman or early human system of communication, but omens of a new and greater awareness.[192] There has, I think, long been a capacity to mix and share experience at a collective level, probably ever since the first complex cells each took strangers into their midst in the form of little lodgers with plans of their own. Now, however, and for the first time ever, there is a species with conscious awareness that this is going on. Perhaps now we are ready to knock holes in the filter system and take great drafts of raw experience instead of the dribbles on which we were raised. Maybe we have grown big enough to handle the shock of being weaned and turned out into the real world.

We seem suddenly to have access to some of the necessary tools.

Paramount amongst these is hypnosis. Stephen Black describes it as "not only the most simple and practical way of proving the existence of the unconscious—which is still in doubt in some circles—but the only way in which unconscious mechanisms can be manipulated under repeatable experimental conditions for purposes of investigation."[58] I concur completely. Hypnotism is a marvelous instrument, which almost anyone can learn to handle in thirty minutes, but there is one still rather basic problem about it—nobody yet knows exactly what it is.

There are no simple cues, like the rapid eye movements of active sleep, to tell us precisely when someone is hypnotized. Sleeping and dreaming can both be differentiated from waking by the differences of the patterns that show up on an electroencephalograph, but the brain waves of a hypnotized person (someone who says afterward that he was hypnotized, or who responds appropriately to a posthypnotic command) are identical with the waking state.[117] A subject wired to an electroencephalograph shows, when resting with the eyes closed, exactly the same pattern of waves as when hypnotized a moment later by a code word.[146] There is no measurable change in cortical potential, pulse rate, skin resistance, or palmar electrical potentials.[332] In short, there is nothing in the physiology or behavior of a hypnotized person that makes it possible for an observer to distinguish someone in this state from someone else who is merely pretending to be under hypnotic control. And yet there is something distinctive and different about being hypnotized which anyone who has had the experience will confirm. It seems to be an altered state, which all subjects learn to recognize in personal internal terms; but as soon as they are asked to define it, irreconcilable differences appear.[484]

Despite decades of intensive research, we are still no closer to understanding hypnosis than Franz Mesmer was when he first began using it in Paris in 1778. It is a phenomenon as elusive as any in the psychic field, with the same familiar and maddening mixture of substance and illusion as telepathy, or metal-bending. This has led many researchers to abandon it, or to deny it altogether, but the fact remains that under certain circumstances, with certain people, it achieves some extraordinary results. For me at least, the very fugitive nature of the subject is a clear indication that it belongs in the area of the unconscious and that, in using it, we are on the right track. I recognize the signs.

It is almost impossible to measure something when you don't know what it is, so it is necessary to try indirect approaches. While the characteristics of hypnosis itself may still be in doubt, we do have a good idea of the nature of the process of hypnotic induction. Procedures are very varied, but as Charles

Tart of the University of California has pointed out, all have certain steps in common.[525]

The first usually involves having you as the subject sit or lie comfortably, so there is no effort needed to maintain your body position, while an instruction is given to relax as much as possible. This has a variety of direct effects. The initial one is to limit anxiety and bodily tension, making it easier for your consciousness to concentrate on itself. And this internalization process is further encouraged as the kinesthetic receptors, which normally tell you what position each of your limbs occupies in space, fade out one by one as the body withdraws from consciousness, taking with it the usual patterns of stimulation that help enforce a sense of wakefulness, and stabilize your state of being conscious.

The second step is for the hypnotist to tell you to concentrate only on his voice or actions, ignoring all other thoughts or sensations that might come to mind. In normal wakefulness, you scan your environment constantly, ever alert for information and stimulus. This pattern of attention is what enables you to pick meaningful signals out of the confusion of environmental noise, discriminate between rival patterns of information, decide which ones are important, compare them with memories in store, and prepare to make appropriate response. But the moment attention and awareness are withdrawn, or the normal environment is restricted in some way, the whole chain of reaction begins to attenuate and finally breaks down altogether, removing a major part of your usual wakeful psychological load. The most effective mechanical device for inducing hypnosis without any vocal instruction is an apparatus that transmits magnified breathing sounds from a microphone on the lower throat to earphones that cut out all other extraneous noise.[329]

A third common step is an instruction not to think about what the hypnotist is saying, but to listen passively. If he says your arm is feeling heavy, just accept that. In an ordinary conscious state, you think constantly about everything that is said to you, evaluating it and making decisions about meaning and content, activating other mental and emotional subsystems. Interruption of this sequence removes part of the usual load of activity that helps maintain waking attention, and even further undermines consciousness.

Next, the hypnotist might tell you, if you still show signs of restlessness, to look fixedly at some single object. Normally, this is something you never do, because it produces retinal fatigue. And when you do try it, under instruction, all sorts of unexpected visual effects occur. Colored halos materialize around the object, shadows appear and disappear, portions of the object fade. Because these are not part of your normal experience, they disrupt processing centers in the brain and help to enhance the prestige of the hypnotist, who seems to have the power to do these things to you.

Then comes the critical step. The hypnotist suggests that perhaps you might feel a little drowsy or even inclined to sleep. Just the mention of the word "sleep" elicits memories of sleeping, and even if you resist actually going to sleep, your consciousness has been disrupted again, passivity is reinforced, and your body image tends to fade even further. But, at the same time, the hypnotist might tell you that what you are going to experience is not true sleep because you will still be able to hear him. In practice, in our culture, this precaution is seldom necessary because most of us know what to expect of the hypnotic state, and are familiar with the idea that it involves a sleeplike condition in which we are still able to respond to direct suggestion.

Finally the hypnotist clinches the agreement you and he have made by offering several simple suggestions which have built-in physiological responses. He asks you to raise your arm and suggests that it feels heavy. Of course it does, it is heavy, and as you appreciate this, his prestige is magnified yet again. This strikes directly at your sense of identity, because normally it is your own inner voice that tells you what to do. Now the hypnotist's voice is taking over this role and your sense of self has to expand to include him as well. He goes on from simple motor suggestions to more and more complex actions and ideas, and with each new success, your ego boundaries become even further blurred, and you slide into the trance state we know as deep hypnosis.

The basic effect of the whole induction procedure is to weaken the boundaries we normally erect between "me" and "not-me," enabling hypnotist and subject to engulf each other and to become totally identified one with the other. Lawrence Kubie of the University of Maryland says that "when the demarcation becomes unclear between that which seems to be 'me' and that which seems to be 'you,' it is hard to decide who originates any words, criticisms, praises, commands, or suggestions."[328] Under hypnosis, we even stop trying to make these kinds of distinctions. We lose identity and integrity and revert, at least in part, to an earlier evolutionary condition, in which we mixed and shared more freely with our environment.

I have already shown how individuality seems to be the result of a complex organization of a certain kind; and how this ordered association came about because each component of the system had learned, at an even earlier stage in evolution, to exercise selective sensitivity, discriminating between "me" and "not-me." The major problem facing the first organisms was the basic recognition of the material substance of "self" from the material substance of "non-self" in the environment. This evolved, once organized communities reached the complexity of vertebrates, into an immune system which put up special defenses against invasion. Every time an antigen provoked an owner of an immune system, it responded by producing an antibody which was stored and maintained in case it should ever be needed again. It provided, in other words, a very early physical form of memory. It was, I suggest, precisely this mechanism that forced early nerve nets to specialize in ways that have produced our central nervous system. Our mind has its origins in the early immune response, in the capacity to say "this is me" and "that isn't." And the compartments in the mind, the boundaries between consciousness and the unconscious, must therefore be maintained by the psychological equivalent of an immune response. Our mental censor, the filter, seems to be a system which distinguishes between "me" and "not-me," and if that is so, then the best way around it will be to abolish that distinction. Which is exactly what hypnosis does.

If there is anything in this theory, then one of the direct effects of passing through the process of hypnotic induction should be a capacity to bring the immune response under conscious control.

One of the visible signs of an allergic reaction is the production of blisters on the skin. A blister, or to put it in precise medical terms, an ecchymosis, occurs when there is an increase in permeability of the walls of small capillaries that allows histamine to flow into tissue, where it produces inflammation, dilation of the arterioles, and the characteristic mottled swelling. Frank Pattie of the Rice Institute discovered one patient, a thirty-eight-year-old woman, who was able to produce such blisters in direct response to nothing but the hypnotic suggestion that an allergen had come in contact with her skin. In one carefully controlled test, designed to eliminate all possibility of her deliberately injuring herself at the chosen site, her arm was covered in a plaster cast with a watch glass set into it; and when it was suggested to her that the area directly beneath the window was affected, she produced a blister there within five minutes.[425]

In an elegant study in Japan, a number of subjects known to suffer from allergic dermatitis when exposed to the leaves of a local tree were blindfolded and tested both with the leaves of the offending tree and leaves from a harmless common chestnut.[276] Under hypnosis, all of them produced blisters when touched with chestnut and told it was the allergen; and none of them reacted to prolonged contact with the usual source of their ills. It was also discovered that a second group of allergic subjects were able to demonstrate a degree of false-blistering and a certain amount of protection from the guilty tree, even though they had not been hypnotized. We are, it seems, capable, given the right attitudes, of exercising conscious control of the "me versus not-me" reaction, but contrive to do this much better when we have been persuaded by the hypnotic induction process that "self" is being held in abeyance, and boundaries abolished.

It seems likely, in the light of experiments of this sort, that control of psychosomatic ills such as ulcers, colitis, migraine, eczema, and high blood pressure does not depend on taking control of one's own body in some Dale Carnegie, positive-thinking sense, but may best be accomplished by relinquishing consciousness altogether.

We may, as a species, have a propensity for doing this anyway. Something which we learn during the long period of dependence we undergo as infants, when we have to take so much on trust. We even, it seems, take orders in our sleep. In one sleep laboratory, experimenters tried whispering instructions such as "Whenever I say the word 'blanket,' you will feel cold until you pull up the blanket and cover yourself." Or "Whenever I say the word 'itch,' your nose will feel itchy until you scratch it." Electroencephalograph recordings showed no interruption of the brain patterns of quiet sleep, and the subjects knew nothing of the instruction on waking. But the following night, all responded appropriately to the signals. One even continued to scratch on cue five months later without any reinforcement during the interval.[167]

This looks to me like a mechanism designed to ameliorate the dangers of sleeping—a sort of watchdog, like the dream, which stands guard over us in times of vulnerability, ready to respond to even a whispered warning. If it is, then one would expect certain words or sounds to be more effective than others. Ian Oswald of the University of Edinburgh tested the ability of the sleeping brain to distinguish between different signals by playing a very long tape recording of fifty-six names called out one after the other in various orders, with several seconds between each name.[416] He found that the electro-encephalographic recording of slow-wave sleep signals was most disturbed by the subject's own name, or that of someone important to them. "While Neville slept, the name of his recently acquired heart-throb, Penelope, would cause a most violent perturbation in his EEG, and a huge psychogalvanic response, or sudden sweating of the palm." Proof that it was the name itself, with its associations, which had this effect was provided by playing the tape backward. In this reversed mode no meaningless sound had more effect than any other.

This similarity of response in quiet sleep and under hypnosis doesn't mean that the two states have anything more in common than the fact that both show a diminution of waking consciousness, with a corresponding increase in the activity of unconscious areas. There may, however, be a closer connection between dreaming and hypnosis in that both seem to be able to draw freely on material not normally available to consciousness.[166]

Under hypnosis, most people seem to be able to recall almost everything that ever happened to them. I know that in one session I was able to conjure up an image of the route I used to take while walking to school at the age of six in Johannesburg, and count all the street lights on the way. Two years ago, I returned there and for the first time in thirty years, retraced my steps, and was able to verify a count I don't think I ever consciously made. This ability to relive early experience in extraordinary detail has made hypnosis a useful tool in psychotherapy, because any analysis boils down in the end to a doctor-versus-patient battle, with the patient trying to retain his neurotic compulsions while the practitioner does his level best to eliminate them. Some therapies involve role-playing and the relief of tension by acting out traumas, but nothing is quite as beneficial as full recall of the actual experience.[361]

This is usually produced by the therapist creating confusion as to the present date, and then talking the subject back down through the years, step by step, stopping at regular intervals to establish ground with comments like "You are now ten years old. Where do you go to school? What are you wearing?" The replies are often fascinating, showing language, logical constructs, and personality traits appropriate for an individual of the relevant age.[199] Sometimes the subject is asked to write something down, and as they apparently travel back through time, the script actually takes on more of the characteristics of a childish scrawl. In several regressions I have seen subjects who have been taken back to a time when they went through a change of handwriting revert once again to a style they hadn't used for decades.

Arguments still rage, however, about the validity of the age-regression technique. Those who maintain that reenactment of early states is no more than half-conscious dramatization point to the fact that an American adult regressed to a childhood in which he spoke only German continues nevertheless to respond to questions asked in English. Or to the fact that a subject supposedly regressed to infancy continues nevertheless to sit upright in her chair in adult fashion.[33] There is no question that part at least of the personality remains at the present level and functions as a sort of detached observer, but there is equally little doubt that a great amount of actual revivification is taking place.

Robert True of the University of Vermont took fifty students known to be deep hypnotic subjects, between the ages of twenty and twenty-four, and regressed them to the apparent ages of ten, seven, and four years old.[547] At each of these points he asked them to pause on Christmas Day and on their birthdays, and tell him which day of the week it was. Then he checked their replies against a permanent calendar and found that at the age of ten, ninety-three percent of the answers were correct. At the ages of seven and four the figure fell to forty-one and thirty-five percent respectively, which is still far greater than the one-in-seven chance of making a correct guess at random, and probably reflects the fact that younger children are progressively less likely to have known the day in the first place. This experiment has been criticized on the grounds that "within a few seconds one can mentally compute the day of the week on which an earlier birthday fell, by counting back from a known birthday,"[31] but I find that rather lame. You try it. I can't even remember the day on which my birthday fell this year. And even if it were possible to cheat in this way, why not get as many correct answers for the lower ages, which don't involve a calculation any more difficult, or any easier, to make?

An even more conclusive test of the validity of regression would be evidence that the subjects showed involuntary behavioral patterns they now no longer possess. Babies are bundles of strange reflexes, some of which are extremely short-lived and seem to appear only for long enough to pay temporary lip service to their evolutionary past. Two of the best known are the rooting reflex, which makes a baby turn its head toward a touch on the cheek or the corner of the mouth, seeking out the nipple with sucking movements; and the Babinski reflex, in which the toes tend to turn upward and spread apart when the sole is stroked. Rooting stops after the first year and the Babinski reflex is replaced between six months and two years by the persistent adult pattern of turning down the toes.[491]

Two researchers at the Columbia Medical Center in New York tried taking subjects back to the first year of life and found that some of them curled up into an infantile sleeping posture, head on one side and fists clenched and held up near the face, and when touched on cheek or foot, they produced the full rooting and Babinski responses. When progressed again beyond the critical age of two years, the subjects reverted to their normal adult patterns.[201] Not unexpectedly, this experiment has also come under heavy fire. The most critical comments point out that "extreme relaxation or depressed muscle tone is sufficient by itself to produce a Babinski response."[32] Perhaps; though it is only fair to point out that this is not a full Babinski, which in adults is reserved for those who have tetanus, strychnine poisoning, epilepsy, or permanent damage to the pyramidal tract. And none of the critics has been able to suggest why the subjects should necessarily have reached this critical state of relaxation only when regressed to the age of five months or less. Almost invariably, the spoilers end by falling back on allegations of possible fraud, to which unfortunately there is no valid response. For these super-rationalists, it is necessary to reject all evidence for the paranormal as mistaken or false. They examine it purely for purposes of refutation, looking for places where impostors could have been at work. "Could have" is then translated into "did," since for them all other explanations are automatically ruled out.

It is true that the design of many experiments could be tightened up to eliminate possibilities of other factors contributing to the results, and it is certain that simulating subjects can produce behavioral responses, and even physiological changes, which are impossible to distinguish from those observed under hypnosis. It seems that good actors, particularly those of the Method School, go through a process of what amounts to autohypnosis before stepping into a role. But, as far as I'm concerned, it doesn't really matter whether the subjects under hypnosis are play-acting or not. This may be what hypnosis is about anyway; we have no evidence to suggest that it is something a hypnotist does to you. However it works, and whoever is responsible, what is important is that the technique of induction gives ready access to the contents and the capacities of the unconscious.

Some subjects who give childlike performances when regressed to childhood are also able to give convincing portrayals of themselves as older individuals when hypnotically progressed to a future time.[316] In one study, five subjects who were "able to relive vividly past experiences" were also able to create futures which "sound possible and well within the realm of probability, as judged from a careful personality study," though they "did not attempt to describe events outside their own lives, except in a most vague fashion."[468] The implication is that, under hypnosis, some people are able to build on and embellish personal knowledge and experience in much the same way as we all do in dreams, which, as everyone knows, can be very convincing. Great care is necessary in determining exactly what is, and what cannot be, made available for imagery by dredging down deep into unconscious areas.

In 1965, a physician in Toronto discovered that one of his patients, under hypnosis, began to speak in a foreign language. When nobody could be found to interpret this, the patient was asked to write some of it down, which he did with great care, printing each character.[147] After much detective work, the fragment of script was identified as Oscan, a language spoken in western Italy up to the first century B.C., when it was replaced by Latin. The only surviving written example of Oscan dates back to the fifth century B.C., and is a ritual curse inscribed on a thin lead scroll which was buried along with a man, presumably to give him power over the demons of the underworld. And it was this phrase, now known as the Curse of Vibia, which the patient had produced, letter-perfect despite the fact that he knew no Latin and had no interest in archaeology. This looked like a good example of information acquired in some unusual way, until a follow-up investigation established that, years previously, the patient had access in a library to a book which contained the passage in Oscan, set out in large letters.[162] He had no conscious awareness of having ever seen it, but it seems that it was imprinted in some way on his memory, and surfaced again only under hypnosis. This may not have been how he acquired and was able to reproduce the curse, but our unconscious certainly does have such a capacity.

Sometimes memories are not recognized as such, but are dressed up in personal detail and experienced as an original production. I have several vivid memories of childhood incidents which my parents insist happened not to me, but to one of my brothers. This confusion is known as cryptomnesia and may well be responsible for much that seems inexplicable in any normal way. Studies on memory abound, but it is another one of those elusive phenomena about which we still know next to nothing. One of the few things about which we can be certain is that recall is not an all-or-nothing experience. At one extreme, we succeed in conjuring up an item of information and recognize it quite clearly as that which we were trying to remember. And at the other extreme, after a period of brain-racking, we feel equally certain that any further effort would be pointless, because we have forgotten. But between these extremes are a range of partial successes, such as the tantalizing tip-of-the-tongue experience, when we feel certain that we almost have the answer, but the harder we try, the more elusive it becomes. Almost all memory studies concentrate on the recognition of external signals—scarcely any attention has been paid to the recognition of signals from within—so the problems in this area remain not only unanswered, but apparently almost totally unconsidered. Nobody knows how we recognize an idea as being old information rather than a new synthesis; or how we are able to assess that the long list of items we run through our minds in an attempt at recall does not include the one we are looking for.[448] So in our current state of ignorance about what a single mind, alone and unaided, can do, it would be rash to assume that any faculty we don't understand must necessarily be the work of an occult agency. And yet, hypnosis brings some incredible things to light, taking subjects back beyond the point of conception to what seems to be other personalities in other situations.

Albert de Rochas, at the turn of the century, was the first in modern times to unearth apparent evidence of a "past life."[128] His hypnotic subject, a young orphan girl with no education, claimed to have been a male clerk in the ministerial offices of Louis XVIII and provided him with masses of historical information which he was later able to verify. The most publicized case of this kind since then has undoubtedly been Morey Bernstein's hypnosis of a young American woman named Virginia Tighe, who seemed to take on the Irish personality of nineteenth-century Bridey Murphy.[53] And very recently, Arnall Bloxham in Britain, who has had comparable success with a number of subjects, experienced a complex hypnotic reenactment of the massacre of Jews in twelfth-century York, as described by a twentieth-century Welsh housewife.[281]

There is a growing body of literature on comparable cases, many of which have been subject to critical analysis and seem to have included information not normally available to those being hypnotized. Bloxham's subject was even able to provide data that were apparently not available at all. In her role as a Jewess in York, she spoke of a crypt beneath what is now St. Mary's church on the site of the medieval city; but no vault was discovered until several years after her hypnosis. Evidence of this kind is being offered by some as proof of reincarnation, which it may well be, but it is perhaps best characterized by the comments of the philosopher Curt Ducasse, who made an exhaustive study of the Bridey Murphy case. Nobody, he said "has succeeded in disproving, or even in establishing a strong case against, the possibility that many of the statements of the Bridey personality are genuinely memories of an earlier life of Virginia Tighe over a century ago in Ireland." But, he added, none of the facts offered in her trance state "prove that Virginia is a reincarnation of Bridey, nor do they establish a particularly strong case for it. They do, on the other hand, constitute fairly strong evidence that, in the hypnotic trances, paranormal knowledge of one or other of several possible kinds concerning these recondite facts of nineteenth century Ireland, became manifest."[138]

That much is beyond question. Some people under hypnosis experience elaborate transferrals of personality with a wealth of corroborative detail. A few, like the novelist Joan Grant, can even achieve this result spontaneously. She uses it to great effect to provide the background for her historical romances.[214] It is even possible, as Renée Haynes suggests, that "Western belief in reincarnation, already adumbrated in early eighteenth century philosophy, probably owes as much to the development of the historical novel as to the teachings brought back from India by administrators and soldiers and scholars, and later by Theosophists."[245] But this fails to explain why a belief in reincarnation should have become so strong in India in the first place; and it avoids the fundamental issue altogether. The whole argument for and against reincarnation is obscuring the important fact that many of us, at least some of the time, have access to amazing information that has no direct connection with our waking lives, and seems to offer little that could be of any immediate importance.

I insist on asking the basic biological questions again and again. What use is this ability? Is there any survival value in it? Why do we do it?

Charles Tart tells of an occasion when he suggested to a deeply hypnotized subject that he was getting up from his chair, going down the hall, and out of the laboratory building onto the University of California campus at Davis. The subject described the whole experience as it was happening to him, telling of being in the yard and seeing a mole pop out of its burrow on the lawn. Tart asked him to catch the mole, and he said he had. After a long mental ramble around the area, he returned in imagination to the building, climbed the stairs, and reentered the experimental room. Here he was asked to pause in the middle of the floor and describe what he could see. He did this in great detail, omitting only any mention of the chair in which he still sat. Then he was asked:

"Is there a chair in the room?"

"Yes."

"Is there anyone sitting in the chair?"

"I am."

"Didn't you just tell me you were standing in the middle of the room?"

"Yes, I am standing in the middle of the room."

"Don't you think it's contradictory to tell me you're standing in the middle of the room and sitting in the chair at the same time?"

"Yes."

"Does the contradiction bother you?"

"No."

"Which one of the two selves is your real self?"

"They are both my real self."

Exasperated by his inability to get the subject to admit that he was hallucinating, Tart paused for a moment and then, in desperation, asked one final question:

"Isn't there any difference at all between the two selves?"

"Yes, the me standing in the middle of the floor has a mole in his hands."[525]

This is a superb example of what has come to be known as "trance logic," which is so common in hypnosis that it is considered by some to be a basic characteristic of the state. If a subject is told that a real chair does not exist, he will no longer see it; but if asked to move around the room, will never bump into "nonexistent" furniture.[412] Clearly, such a person is effectively aware of the presence of the chair at one level, even though he doesn't register its existence at another, more conscious level. He operates as sleepwalkers do, negotiating complex environments without injury, though showing none of the standards of normal wakefulness.[303] When shown a real chair and given the suggestion to hallucinate a person sitting in it, subjects nevertheless acknowledge seeing the back of the chair directly through its imaginary occupant, but this "illogical" state of chair affairs is quite acceptable to them. We are very quick to classify this sort of outlook as a deterioration in logic, forgetting that logic itself is rather artificial, being part of a self-contained structure of assumptions. Thinking in a pattern which contains contradictions according to one system of logic does not necessarily mean that such thinking is invalid or useless.[73]

Trance logic, or as Freud called it, "primary process thinking," in fact embodies the important and fertile idea that an organism can register information at one level of functioning that isn't picked up at other levels. The more we learn about the workings of the mind, the more obvious it becomes that it always can, and perhaps even must, operate on many levels simultaneously.

Take the problem of pain as an example. This is another of the great mysteries of life. It is not a sensation in the part of the body that is hurt; nor is it an event in the nerves that conduct impulses from that part of the brain. It seems to be a process in the brain, but nobody knows quite where, or how it works. Patrick Wall of University College, London, suggests that the process depends not only on an injury, but on existing and past circumstances, and is therefore unique to every individual. "Furthermore," he adds, "if each stimulus changes the pattern of activity by adding memory and recall (in themselves mysterious functions), the code is a shift code which is uncrackable."[568]

It has been common knowledge ever since hypnosis was first discovered that it can act as an analgesic, helping patients to cope with pain and even making it possible, in certain cases, to conduct major open-body surgery without anesthetics. But the mechanism remained mysterious until a minor breakthrough occurred in 1960. A twenty-year-old college student was being tested for sensitivity to pain under hypnosis, by having his left hand pricked with pins, following a suggestion that this hand was anesthetized. He seemed oblivious to what was happening and claimed to feel no pain at all; but his right hand, which happened to be holding a pen at the time, began scribbling frantically, "Ouch, damn it, you're hurting me."[305] Pain clearly was being registered at some level.

Ernest Hilgard at Stanford University suggests that we all have what he calls a "hidden observer," who takes part in a secret internal system of communication.[251] Subjects in his tests put one arm into circulating iced water, which cuts down on local circulation, and within a couple of minutes produces normally unbearable ischemic pain. Under hypnotic analgesia, they seem to experience no discomfort, and claim to be free of pain even when the experiment goes on for long periods; but all the subjects also receive an additional hypnotic suggestion: "When I place my hand on your shoulder, I shall be able to talk to a hidden part of you that knows things going on in your body, things that are unknown to the part of you to which I am now talking." And, when touched on the shoulder, they all report almost as much pain and suffering as subjects who have not been hypnotized at all.[318]

It is important to appreciate that this admission of pain by the hidden observer doesn't disqualify the subject's honest denial of pain at another level. It is an anomaly that highlights the complexity of the response, taking it beyond the simple-minded idea that you either have pain or you don't. It is possible to have both experiences at the same time. It is possible to see a chair, and simultaneously not to see it. It is possible to hear a word whispered to you in your sleep, and not to hear it at all. It is undeniably possible for you to sleep soundly and yet still retain sufficient awareness of your surroundings to prevent your nocturnal turnings from carrying you over the edge. Most adults just don't fall out of bed. Perception is more than a single all-or-nothing response. Trance logic is valid, and is not limited to hypnosis. It is typical of dreams and other dissociated states, and in all of them it makes excellent biological sense.

When we first acquired consciousness, we gained the most powerful instrument ever to be put at the disposal of evolution, but we also put ourselves at risk. The danger in reflection is that it is so fascinating it tends to dominate everything else, leaving us open to outside threats. Kenneth Bowers of the University of Waterloo in Canada tells of a talented opera singer who finds, on the night of particularly powerful performances, that she sometimes surfaces in a panic with "no clear memory of having sung an aria that, according to the ongoing music, she must have just completed."[73] It is as though she becomes so absorbed by her singing that even the sound of her own voice is dissociated. Even the usual hidden observer is deeply involved, and there is nothing left of "her" to notice that "she" is singing.

Legendary "absentminded" professors operate in the same way, leaving wives and colleagues to pick up after them and take care of the little everyday details of survival. In our culture, we tolerate such idiosyncrasy because we value the output of these specialized people, but this was certainly not true when consciousness first began. It was every organism for itself and the devil, or a dinosaur, take the hindmost ones too singleminded to react to warning signals.

In the beginning there were only automatic responses; everything was unconscious. Then, when a critical level of complexity in the community of cells was reached, there were the first glimmerings of true conscious awareness, bringing both vigor and vulnerability. The new talent was worth keeping, but the organization had to be protected, so right from the start it was necessary to seal consciousness off in its own cerebral compartment, creating a two-tier system with a censor or filter in between.

At first, the filter may have been permeable in only one direction, passing information acquired by the unconscious up to conscious levels when it was significant enough to warrant such a breach of protocol. But there was also survival value in relegating unbearable parts of the conscious content, and the barrier became permeable in both directions, creating the first personal unconscious as a sort of garbage dump.

There is obvious adaptive value in this arrangement. Being in two minds simultaneously means being able to process information continuously without interruption to the conscious stream. A capacity to collect data without having to think about it, beyond the bounds of awareness, must offer a species considerable advantage. If we were forced to pay conscious attention to our environment, we would be in a permanent state of distraction, unable to string even two simple thoughts together. The power to think at all depends entirely on an ability to register background information at a relatively low level of analysis, bringing it up to consciousness only when it carries vital information, like the snap of a twig in the forest, or the smell of something burning in the kitchen.

Without perfection of this split-level system, we would almost certainly be totally incapable of any of the special skills, such as driving or playing tennis, which, once they have been mastered, become largely automatic. The leaders in these areas seem to behave like operatic divas and absentminded professors, putting everything, conscious and unconscious, into their performances. But for casual drivers or weekend tennis players, nothing is more detrimental to skill than the kind of Stephen Potter ploy which results in deliberate conscious interference with normally effortless and totally automatic gear changes or backhand swings.[439]

Dissociation is vital. Under normal circumstances the hidden observer must have no more than a watching brief, acting as a guardian angel, giving consciousness the chance to develop and grow. When consciousness is diluted, as it usually is in sleep, it is the observer that throws up the occasional dream to keep the system as a whole on its toes. In fact, the dream stream itself may serve the additional and indispensable function of reinforcing dissociative abilities, giving us each the chance to practice trance logic every day of our lives. There seems to be a direct link between intelligence and the ability to dissociate, to cut mental corners and practice unusual creative, lateral thinking. All invention probably depends on a delicate interplay of liberal dissociation and more conservative traits. All truly creative insight seems to flow from breaches in the barrier. It can't be purely accidental that Coleridge composed *Kubla Khan* in his sleep; that Mozart found his best musical inspiration rising like dreams, quite independent of his will; and that Friedrick Kekulé made his revolutionary perception of the cyclic structure of carbon rings while sitting in a trance on a bus.

We need both conscious and unconscious mechanisms, but we perhaps have an even greater need right now of interconnection between the two. Absentminded professors do it naturally and relatively harmlessly; and hypnosis achieves similar results experimentally, showing that each of us has at least the potential of becoming a Nikola Tesla. We all have access to genius. The process involves a new diminution of voluntariness and a deliberate adoption of trance logic; both, as it happens, characteristics of meditation and many of the popular new techniques of transcendence. I believe they have become growth industries in direct response to an evolutionary pressure in just this direction. Few of them have any real survival value, and some are downright dangerous; but I suspect that even the systems which are harmful may not be doing much more than accentuating existing pathologies.

Improper dissociation results in fugue states, in which a person wanders off not knowing who or where he is; in traumatic neuroses, produced most often by unexpected experiences which a person is unable to assimilate properly, either by coping with them consciously or by filing them away in the personal unconscious; and in multiple personalities.[202]

Pierre Janet, the famous French psychologist, believed that hypnosis itself was a condition of dissociation in which part of consciousness broke away, forming a secondary personality. One of his subjects, known usually as Leonie, became under hypnosis a much more animated character who called herself Leontine. Initially the two were distinct, but at one point Janet received in the same envelope separate letters, each in its own style and handwriting, from both personalities, who were on holiday together. Later, in a session under hypnosis, Leontine suddenly stopped her usual noisy monologue and said, "Oh, who is talking to me like that? It frightens me . . . I hear on the left a voice which repeats, 'Enough! enough! be quiet; you are a nuisance.' " "Assuredly," concludes Janet, "the voice which thus spoke was a reasonable one, for Leontine was insupportable; but I had suggested nothing of the kind and had had no idea of inspiring hallucination of hearing." He christened the new voice, which seems to have been that of the unconscious hidden observer, Leonore. And he found that this personality had a capacity for complete detachment in which the organism which housed the three ladies enjoyed ecstatic experiences. "She grows pale, she ceases to speak or hear, her eyes, though still shut, are turned heavenwards, her mouth smiles, and her face takes an expression of beatitude."[389] It was in this condition that the Leonie-Leontine-Leonore complex produced most of Janet's best telepathic material.

An even better-documented case is the treatment by Morton Prince, an American psychologist, of a student nurse known as Christine Beauchamp who had four quite distinct personalities.[442] Prince simply labeled these B1, B2, B3, and B4. The most frequent state was B1, who was quite unaware of the others. B2 was quiet and mature and recognized B1, but not B3 or B4. The most problematic was B3: "A rather childlike personality, she was utterly irresponsible, devoid apparently of all conscience and lacking moral scruples." B3 would go to any lengths to upset B1, posting her boxes of live spiders, unraveling her knitting, hiding her money, making her lie awake and throwing all her clothes on the floor. When Prince tried to eliminate B3, she wrote him letters asking him to discontinue treatment of B1 and threatening dire consequences. B4 appeared suddenly late in the treatment and showed "far more strength of character, albeit irritable, headstrong and quick tempered," and almost immediately began a devastating battle with B3 for control of the complex. The turmoil only ended when Prince discovered that if he hypnotized B4, then B2 appeared, retaining all the memories of B1 and B4. He concluded that B2 was in fact the real Miss Beauchamp. He suggested to B2 that when awake as B1 she would remember all about B4, and when awake as B4 she would think and feel in the manner of B1. By repeatedly hypnotizing and awakening her, alternately as B1 and B4, these two personalities blended and faded into the background of B2, who emerged triumphant. B3 at this point vanished without trace.

Recently both Leonie and Christine have been topped by a patient of Cornelia Wilbur, a New York psychoanalyst, who seemed to have sixteen distinct personalities.[478] These appeared gradually over a period of twenty years and even included two male characters. After an intensive analysis of all the "people" involved, they were successfully subsumed in a new seventeenth self who became a stable amalgam of all their characteristics. "Do you know," this one asked, "what it means to have a *whole* day ahead of you, a day you can call your own?" Instead of having to be content with a mere sixteenth.

The formation of a multiple personality is still mysterious, though all the indications suggest that separate identities are taken on to make it possible to escape from oppressive situations. They act out different and incompatible needs. This escape clearly is a strategy of the unconscious which results in the fragmentation of a single consciousness. In their separate state, the divided selves coexist. When one is in control, all the others are temporarily dissociated, much as the other independent gears are inoperative when you happen to be driving in third. All the personalities involved, however, can be tracked down to the same source. None is in any way external, and there is no question of possession. But the separate identities have an extraordinary autonomy. Each has its own separate memory bank recalling mainly those things which happened during its turn at the controls, like the millionaire in Chaplin's classic *City Lights*, who recognizes Charlie only when he is drunk.[211] There is seldom any overlap between the personalities, and even under intensive analysis it is almost impossible to detect any leakage between them. In all the psychological tests that can be devised, they react like different people. In one study of a twenty-seven-year-old black American with four separate selves, even their electroencephalograms were different.[355]

This man, Jonah, had split under the cumulative effect of a number of traumatic experiences, adding to his usual quiet conservative self three others: a warrior, who acted out all his anger, brawling in bars and smashing up his home; a lover, who took care of his sex life; and a mediator, a lawyer-type, who went to great pains to patch over all his interpersonal conflicts. All our personalities are shaped by a variety of needs, and though one or another may prove dominant or more intense at any given time, there are no clear boundaries between them. Each is integrated, along with all the others, under the umbrella of our total individuality. But for Jonah, each character had its own separate historical development and was walled off from the others. External conflict simply activated a switch from one to another, leaving each with its own inventory, memory, intelligence quotient, and characteristic brain waves.

Role-playing of this order is one of the most impressive talents of the unconscious. Given little more than the genetic outline of a script, it draws freely on environmental information and experience to flesh out the part it is called on to play, and doesn't hesitate, when presented with any serious conflict of interests, to rewrite the role altogether for two or more actors. Multiple personality problems may be a lot more common than we realize. Jonah was seen by no less than thirteen psychiatrists before he was lucky enough to hit on one familiar with the symptoms. This one handed him over to a group, but even then the help they were able to give him was limited. After long and careful analysis, the group was able to persuade his four separate selves to join forces in one unstable, unholy alliance. But the therapists concluded that this new identity was probably psychiatrically "sicker" than any of the others. "This would imply," they say, "that the separate functioning of alter identities may represent a more effective way of handling anxiety than a coalescence of identities."[355] It is just possible that, under certain circumstances, four heads are better than one.

All of us, no matter how well adjusted we may be, have a touch of multiple personality. But in the normal course of events, the usual temporary dissociations of dream, daydream, and reverie are sufficient to ease our growing individuality into a comfortable accommodation with local custom, current fashions in logic, and the prevailing consensus about reality. Given that all these things are purely arbitrary anyway, that is a considerable feat.

Our view of the world is governed by our perception of it. And perception, as I have already been at some pains to point out, is not passive reception or matching; it involves active structuring and synthesis. The result is a vivid, highly textured pattern, very rich in detail. But it depends on surprisingly few inner processes; just a handful of chemicals and an on-off response from each nerve cell. The ultimate possibilities are limitless, but there tends to be a large degree of similarity in the early stages. This may be why most of us, seventy percent according to a recent survey, have frequent feelings of *déjà vu*;[393] of fleeting recognition; a sense that all this has happened before. It is possible that, in this sense of familiarity, it is not that we recognize the features of the stimulus situation itself, but that we have carried out a similar perceptual process before. It may be an internal rather than an external recognition.

On balance, I am so impressed by the glimpse hypnosis gives us of the intricacies of mind and the vast, apparently limitless, capacity of the unconscious to collect and structure information that I feel no need to posit unearthly elements in order to account for seemingly supernatural events. There is enough in genetic memory, in pre- and postnatal experience, and in our astonishing unconscious creativity to deal comfortably with almost all cases suggestive of alien intervention, spirit possession, demonic obsession, and reincarnation.

And having used hypnosis to pry open the door to this secret garden, I believe we can now begin to make our first faltering steps in there on our own.

# CHAPTER ELEVEN

# INVOCATION:
## Experiencing the Present

There are human vampires. Not only in Transylvanian night-mares, but in the person of everyone of woman born.

For the first nine months of our lives, we exist in what one textbook defines as a "parabiotic union between two different organisms in which not only is there intimate apposition and commingling of tissue cells of dissimilar genetic makeup, but also a chronic, covert exchange of blood."[420] A rather long-winded way of saying "pregnancy," but one which effectively stresses the strange nature of the mother-fetus relationship. By all standards, the embryo is a foreign substance and ought to be rejected by the mother's immune system. She is, after all, a vertebrate with a fully functional antibody mechanism that is specifically designed to discriminate "me" from "not-me." But, in some still mysterious way, Nature successfully violates all the laws of transplantation, holding them in astonishing abeyance for the two hundred and seventy-odd days we all spend as parasites in the bodies of our maternal host.

From the gene's point of view, it might be more precise to describe the relationship as symbiotic, because the mother has a vested interest in her offspring's survival. Half of its genes are her own. In purely biological terms, the association of mother and infant in this way is the next logical step in my series of increasingly complex models. I have already said that communities gather because there is survival value in their being together, cooperating in some way, and that novel prop-erties and capacities arise almost accidentally out of such as-sociations, in which the total turns out to be very much more than just the sum of the separate parts. I stress again the notion that form has characteristics of its own, quite independent of its components, and suggest that the mother-child symbiosis has some fascinating possibilities.

Jan Ehrenwald of the Roosevelt Hospital in New York calls pregnancy "the cradle of ESP."[149] He suggests that the inti-mate physiological association of mother and fetus might extend into psychological areas, allowing their egos to merge in a way that bridges the usual gap between individuals. There is certainly no other time in our lives when we get that close to another person, and it would be surprising if this didn't leave some lasting mark. It could in fact pervade almost every aspect of our lives, not only shaping our characters, but providing the basis for a fundamental biological understanding of otherwise inexplicable things.

At the University of Houston, rats were trained by a series of electrical shocks to be afraid of the dark. A substance known as scotophobin was later extracted from their brains and injected into normal mice, who seemed as a result of this chemical transfer also to have a tendency to avoid dark places.[553] There is just a remote possibility that memory can be transferred in this way, perhaps even across the formidable placental barrier, but if there is to be any complex communication between the partners in a mother-fetus symbiotic link, it will have to be of a nonchemical kind. It could be significant that the uterine conditions are almost perfect for hypnotic induction. Temperature and light are virtually constant; the fetus floats at ease in the amniotic fluid, free to drift as it will; and the loudest sound around is the regular, metronomic thud of the maternal heart. Dissociation is almost inevitable, and if the little we have learned about conditions conducive to unusual perception is true, telepathy is more likely to occur in this situation than at any other time in life.[524] And, if this is where the capacity first appears, it will always carry the imprint of that early stage of development.

There is almost no limit to the fertility of this symbiotic theory.

Right away it begins to make sense of the childlike and irrational qualities of so many paranormal phenomena. Most mediumistic communications are repetitive and seem often to be almost simpleminded. Psychoanalytically they would be classed as regressive, which is exactly what can be expected if we are dealing with something that has its roots in infancy. It begins to make sense of the predominant trance logic and primary-process thinking of all trance states, because this is almost certainly the only way a permanently dreaming embryo will be able to think about anything at all. It explains why so many so-called extrasensory perceptions are of a preconceptual, preverbal nature, and therefore almost impossible to describe. It certainly accounts for the fact that spontaneous adult telepathic experiences occur most often between mothers and young children. And it becomes easier, with the help of this model, to understand why all psychic experience is much more common in young children, who have not yet reached the stage of sharply delineating their own ego boundaries, of erecting barriers to defend their mental processes from those of their mothers.

Embryonic origins may be behind the fact that many analysts, beginning with Freud himself, occasionally experience some sort of telepathic link with their patients.[129] The good ones perhaps have such experience more often than others, because the classic psychoanalytic situation is structured to permit regression in the interests of treatment, and is ideally suited to the re-creation and recapture of early experience, such as the first symbiosis. And embryological conditioning could go some way toward explaining why there is such a strong tendency among some people to deny telepathy altogether, even without examining it at all; perhaps because it embodies the threat of what Freud called "the return of the repressed."[189]

If a growing child, by progressively defining the edges of his own ego, smothers the possibility of telepathic communication, then it is also possible that those individuals who fail to establish independent egos may remain vulnerable to continued telepathic intrusions. Most auditory hallucinations are sensed as coming from multiple voices, and where they say anything intelligible, the comments are usually anonymous and expressed in the second or third person.[350] They are always sensed as separate from the self and out of the listener's control.[221] There certainly is an embarrassing similarity between the paranoid schizophrenic's delusions of persecution, his conviction that others can influence his thoughts from a distance, and the theories of those, like myself, who believe it is possible for there to be contact between minds without any normal physical agency. Perhaps schizophrenics really do hear voices.[150] They are definitely selectively attuned to subliminal, repressed hostility in other people, but maybe there is more to it than that. There may be such a thing as "psi pollution" for those with imperfect ego defenses.[152] They may remain open, not only to their mothers, but to everybody else's mother too.

This is an appalling idea, at the moment impossible to prove, but it contains the germ of something that feels familiar, and might just be on the right track. We may all continue, at an unconscious level, to pick up information from foreign sources, information without direct survival value, of so little significance that it stands no chance of breaching the barrier and being made available to consciousness unless special techniques are used to invoke it. Hypnosis is certainly one of these, but there are others.

Every culture has at some time devised a way of circumventing the cerebral censor and communicating directly with the unconscious, usually by pretending that the information is coming from somewhere else, most often through some kind of scapegoat. In Tanzania the Safwa blame a special chair which stands still or shakes in response to questions from a sitter.[243] The Nyoro people of Uganda use a length of wood, something like a spear shaft, called a *segeto*.[42] This is moistened, usually with the blood of a freshly slaughtered goat (the unconscious thrives on that sort of vivid imagery), and the questioner grasps the shaft with his finger and thumb, running them up and down. The point where they stick indicates the oracle's answer. Among the Zande of the Upper Nile, the preferred technique is the *iwa* or rubbing board, which is maneuvered over the flat surface of a special table, answering questions according to whether it slides or sticks.[169] In Europe and the United States in the mid-nineteenth century, turning tables was practically epidemic. "When a group of people sat around a table, touching their fingers to its surface and concentrating on the movement the table should make, it was rare indeed that their thoughts were not carried out."[109] Spirits were soon held to be responsible and codes were devised to communicate with them. Then, as interest in table tipping waned, the planchette came into its own. This was a miniature table, just six inches long and shaped like a heart on wheels, with a pencil fastened to it so that answers could be written down directly. And in 1892, a cabinetmaker in Baltimore patented a lettered board incorporating a small planchette-like indicator, under a process identified by the combined French and German affirmatives as *oui ja*. He made a fortune.

Given these freedoms to express itself anonymously, the unconscious fairly babbles away. In St. Louis in 1913, one particular ouija board began an extraordinary monologue that only ended twenty-five years and four million words later when its operator, a housewife called Pearl Curran, died. Proverbs, poems, and stories poured out of the board in biweekly sessions, picking up the thread each time just where the last effort ended, as though there had been no interruption. And each of the communications was signed with the name of Patience Worth, who was, by the board's own admission, beginning her literary career almost three centuries after her death at the hands of hostile Indians, soon after arriving in the

New World from her home in England. The content is slightly cloying, in the style that *Reader's Digest* likes to describe as "inspirational," but it is extraordinarily consistent. After five thousand poems and several best-selling novels, one is left with a vivid impression of a woman, voluble and often witty, fond of finery and simple pleasures, who might well have lived in the seventeenth century. Most of the compositions are couched in a sort of archaic language which, in the words of an Elizabethan scholar, "is not that of any historical age or period; but, where it is not the current English of the part of the United States in which Mrs. Curran lives, it is a distortion born of superficial acquaintance with poetry and a species of would-be Scottish dialect."[283]

There is no evidence that a Patience Worth ever existed, and suspicion must fall heavily on Pearl Curran, who found, after some years, that she was able to call out the words faster than the board could spell them. Later she graduated to transcribing directly onto a typewriter, finishing one novel set in medieval England in just thirty-five hours. But the psychic researcher Walter Prince, after a year spent observing the situation, concluded that it was impossible for a woman of her limited interests, education, and resources to have written even a small part of the avalanche of words and whimsy. "Either our concept of what we call the subconscious must be radically altered, so as to include potencies of which we hitherto have had no knowledge, or else some cause operating through, but not originating in, the mind of Mrs. Curran must be acknowledged."[443]

Now, a half-century later, we know a little more about the capacity of the unconscious, which seems to be limitless. We also know that it is possible for the unconscious to be in some way directly aware of what is happening to other people beyond the reach of the senses. Charles Tart has shown that when subjects are asked to guess when a friend in a distant room is receiving a random but painful electric shock, their conscious response is totally inaccurate. But if they are monitored for unconscious physiological response, their electro-encephalographic patterns show that at the precise instant of each shock, they react as though they themselves were receiving a mild sensory stimulus.[523]

I think the evidence we already have is overwhelmingly in favor of spontaneous, and at least occasional, reception of telepathic information. It might even be sustained throughout our lives, though it is hard to imagine the nature of a filter capable of sorting out anything relevant from the noise of more than three billion transmitters. As Tart points out, "It is preposterous to assume that only the one person in the world who has been designated as a sender is putting out whatever energy conveys ESP at that particular time; and yet the message gets through, as successful experiments have shown."[526]

Knowing this, and knowing something of the creative skill exercised by the unconscious in constructing internally consistent personalities, I see no major difficulties in the assumption that Pearl Curran could have been Patience Worth.

In December 1963, Jane Roberts and her husband sat down in their apartment in Elmira, New York, to experiment with a Ouija board for the very first time. In the initial two sessions little happened, but in the third they began to get highly articulate answers to their questions. Almost immediately Jane realized that she was receiving this information in her head and could easily vocalize it without the use of the board. So began her communication with a personality which called itself Seth and which has now dictated over five thousand pages of didactic information, organized into informal lectures on subjects such as health, dreams, astral projection, reincarnation, and analytical psychology.[454] The presentation of all this material is lucid and highly individualistic, embodying a wide knowledge of esoteric teachings to which Jane seems to have had no normal access.[455] Eugene Barnard of North Carolina State University explored Seth's character in sessions with Jane when she seemed to be speaking for Seth, and concluded that he had been in conversation "with a personality or intelligence or what have you, whose wit, intellect, and reservoir of knowledge far exceeded my own . . . In any sense in which a psychologist of the Western scientific tradition would understand the phrase, I do not believe that Jane Roberts and Seth are the same person, or the same personality, or different facets of the same personality."[453]

I wish I could be that certain. I have a sneaking feeling that all the way through arguments of this kind we persistently underestimate the capacity and the connectedness of the unconscious. We keep on making limited either/or judgments about things. Either, we insist, the material comes directly from Pearl Curran and Jane Roberts, or Patience Worth and Seth are exactly what they claim to be—spirits from the dead. Their control of the two mediums in these cases is only temporary, but I see no reason why Patience and Seth should not be regarded in the same light as alter personalities in a multiple-personality situation. All the evidence seems to indicate that while it is not in action, when it is in effect out of gear, the alternative personality ceases to exist. In a session once with the celebrated medium Eileen Garrett, a psychologist brought Uvani, her spirit control, into a state of confusion simply by asking what he had been doing since their last session.[527] It seems clear that many alternative characters and spirit guides are created for roles that last only as long as they are on stage. Only when they take over completely and abolish the primary personality altogether can one begin to talk about possession. And if they persist, and become permanent, showing all manner of inappropriate behavior, one can begin to think in terms of possible reincarnation.

Ian Stevenson of the University of Virginia has done everything in his power to make the problem of reincarnation scientifically respectable. And now at last, after fifteen years of intensive, almost single-handed, effort, he seems to be gaining some ground. The prestigious *Journal of Nervous and Mental Disease* devoted its entire issue of September 1977 to his work, noting in an explanatory editorial that "Scientists are not dispassionate seekers. Neither they nor their culture can objectively fix the ultimate boundaries of reality or of scientific investigation."[78]

Stevenson defines reincarnation as the survival and subsequent reembodiment of the human personality after death, and points out that personality consists of more than isolated bits of information.[511] To make a personality, the information has to be organized into particular skills. He uses Michael Polanyi's distinction between cognitive knowledge, which is knowing about something, and tacit skill, which is knowing how to do something.[434] And argues from it that we may know all the facts about a skill, but can never learn to use it without actual practice. Therefore skills such as dancing or riding a bicycle or speaking a foreign language are essentially incommunicable and cannot be passed, without actual physical practice, from one person to another by any normal means. So his major research effort is devoted to the discovery of individuals who seem to have acquired such skills spontaneously, which is a perfectly respectable scientific procedure. But the problem has been that in attempting to understand these situations he is, because of their very nature, concerned only with two possible explanations. One is that his subjects are the recipients of extrasensory information, which he doubts; and the other is that they have actually acquired whole or part of some deceased individual's personality. And because neither answer is one which most scientists consider respectable, the question itself is usually assumed to be invalid and regarded with some disdain.

Stevenson's painstaking researches have established more than sixteen hundred cases which he describes as "suggestive of reincarnation." Most are naturally from the Indian subcontinent, Southeast Asia, and the Middle East, where belief in reincarnation is strongest. A typical case of this type begins when a small child, usually between the ages of two and four, begins to remember living another life. Its statements about this life usually harmonize with its behavior in the sense that, if it claims to have been a wealthy person, it is likely to refuse to do menial work, no matter how poor its present family may be. The child often asks to be taken to the places it remembers, and if these can be identified and the journey is made, it is usually found to have been correct in about ninety percent of its statements about the life and surroundings of the person it claims to remember. After five years of age, memories of the past life seem to fade and usually vanish altogether, along with the unusual behavior prompted by them.[507]

A few of these cases satisfy the criteria for reincarnation in that the subjects do indeed not only possess knowledge, but also special and relevant skills. A young Bengali girl produced elaborate songs and dances; an Indian boy began very early to play the classical drums or *tablas* with great skill; and another child showed unusual expertise with marine engines.[508] In two cases Stevenson even found what he calls "responsive xeno-glossy," an ability to speak and respond to an apparently unlearned language.[509] It is tempting to assume that the ac-complishments of all infant prodigies could be explained in this way. Wolfgang Mozart began composing at the age of four; Johann Gauss was correcting his father's mathematics before he was three; John Stuart Mill and Baron Macauley started writing almost before they could walk. All may have been reincarnations. "Unfortunately," admits Stevenson, "to the best of my knowledge, no Western child prodigy has ever claimed to remember a previous life."

He does, however, go on to suggest that the idea of reincar-nation could have "considerable explanatory value for several features of human personality and biology that currently accepted theories do not adequately clarify."[510] Among these he includes childhood sexuality, the origin of homosexuality, early interest in unusual subjects (Schliemann declared his in-tention of excavating Troy before he was eight years old), rejection of parents, strange birthmarks, the differences between otherwise identical twins, and even abnormal appetites during pregnancy.

Stevenson himself admits that "all of the cases I have investigated so far have some flaws, many of them serious ones. Neither any single case nor all of the investigated cases together offer anything like a proof of reincarnation. They provide instead a body of evidence suggestive of reincarnation." And in dismissing the alternative explanation of some kind of assimilation through special sensitivity, he says, "To accommodate authentic cases of the reincarnation type that are rich in detailed statements and in associated unusual behavior shown by the subject, with the hypothesis of super-extra-sensory perception, requires the extension of that hypothesis so that it becomes no more credible than that of survival after death."[511]

I sympathize with his position, but I believe that this conclusion is premature. The gap between the known capacity of the brain and the demonstration of unusual skill narrows with every new discovery in the life sciences. The existence of vast untapped information in the genes, the pressure of alternative memories in the rival systems of every cell, and the growing appreciation of the powers inherent in the unconscious, make it more and more reasonable to assume that even a three-year-old child could, given the right circumstances, inherit or acquire, and then organize, an elaborate second personality. The very scarcity of those with unusual knowledge or skill tends, I suggest, to support this biological explanation rather than that of reincarnation, which, given the sheer abundance of discarnate spirits that ought to be hanging around, queuing up for reembodiment, is astonishingly rare. I think that, on present evidence, the best conclusion is that offered by Stevenson as an intermediate position. "Once considered about as well-defined as an orange by its skin or a tree by its bark, human personalities now appear to be much more extensible and penetrable than they were thought to be. They can invade and be invaded by processes of extrasensory perception. They may even blend together in the manifestation of a different personality that appears to be new, but that in fact may derive from a fusion of the new and the old."[511]

Time and time again, parapsychological research is drawn into a cul-de-sac with a wall at the end marked "Death" and, in one corner, a convenient escape clause in the form of a ladder on which hangs a little sign that says, "This Way for Survival Without the Body." Millions, perhaps even the majority of those now still alive in the world today, believe that survival is possible, and that a subsequent return in some form of reincarnation is likely. They may be right. If they are, we have in this belief a ready-made answer to almost all the remaining problems posed by apparently psychic experiences. But I am suspicious of easy answers. Despite all the extraordinary things I have come across in the last ten years, I find that somewhere deep down in my being, conditioned to think in biological terms, there is a resistance to the idea, a reaction against providential interventions, against convenient *deus ex machina* twists in the plot which resolve all inherent conflicts by revealing that the characters are actually long-lost relatives.

Possession by discarnate spirits might well be possible, but I don't know. I can't know. With our current knowledge, this concept is too complex to prove or disprove; it is too big a jump from what we know to what might or might not be. The most we can say now with any certainty is that awareness, both conscious and unconscious, seems to be determined at least in part by processes that cannot be localized in the brain and might not be physical at all. They may even be sufficiently independent of temporal and spatial constraints to operate beyond the limits of the body, and mix there with others similarly occupied.

One thing that I have learned in the course of my research is to take what people say seriously. They may be, and very often are, wrong in their rationalization or interpretation of experience; but if enough people have a certain kind of experience, or make a particular causal connection, no matter how implausible it might be, I find that there is very often some good basic reason for their belief. I have even been planning a book called *In Praise of Old Wives* in honor of the substance in so many of their tales. And in all the cultures to which I have ever been exposed, there is one recurrent claim which is on the surface so outrageous that it is usually considered hallucinatory or delusional, but which is so common and so widespread and has been recorded for so long that it must be classified as an archetypal delusion. This is the feeling which many have that they can on occasion go out of their bodies.

Some years ago I traveled in Greece with a friend who quite often enjoys out-of-body experiences. She deals with them very matter-of-factly, and describes what she sees in great detail as you or I might recall a vivid dream. But she can do this as it happens, reporting directly back in what seems to be a sort of semihypnotic trance. On the occasion which most impressed me, we had just arrived in Delphi and had spent the morning drinking from the sacred spring, watching the eagles soar high over the cliffs of the Phaedriades, and exploring the ruins of the temple and sanctuary. No matter how many air-conditioned coaches crowd onto the verges of the narrow mountain road, nothing seems to be able to destroy the magic of the place, and it was very much with us as we rested in the shade of the pines above the old theater, and looked out across the valley of the Pleistos, to the blue hills of Arakhova in the distance. I was enjoying the additional distraction of a nuthatch, a small quizzical-looking bird that was hanging head downward from the trunk of a tree making ribald comment about a chain of harvester ants that wound through the rocks nearby, when my friend began to describe a small chapel she could see. It took a while, and several searching looks in the direction she seemed to be facing, for me to realize that she was off again on one of her strange travels.

She described the whitewashed walls, the red-tiled roof, the tower without a bell; she went into great detail about a building next to it of two stories, living rooms over a stable, with access to the upper floor by means of a wide wooden stair on the outside wall; and she pictured both structures standing on a small plateau cut into a mountainside, with a shallow cavern in the cliff at the rear exposing a natural spring and a seat hewn out of the living rock beside it. All this I got like the commentary to a vivid inner film, colored only by her quiet pleasure in the scene; but then something disturbed her and she fell silent and began to twitch as though tearing herself away from something unpleasant, a little like someone waking from a nightmare. She complained of feeling very cold and refused to say what had disturbed her. All I got was the one word, "Blood." It wasn't until we had walked in the sun all the way around the Pythian stadium on the upper slopes of the sanctuary that she regained her color and composure and stopped trembling.

The following morning I set out early on my own to walk from Delphi down into the gorge, through the sea of ancient olive trees on the Sacred Plain, to the port of Itea on the Gulf of Corinth. I left the road above the ruins of the Marmaria and made my way down a mule track that winds in and out of the ravine. It was a perfect, gentle dawn, and I was very happy to be there. About halfway down to the river bed, I came across a wayside shrine. These are common throughout the Balkans and no matter how remote, are usually carefully maintained, often decorated with a candle or fresh flowers. This one was forsaken, unpainted, with a broken glass plate and resident spiders. If I hadn't stopped to look and wonder, I might not have noticed the unkempt path that led off the track down to the right where the top of a tiled roof was just visible through the trees.

I made the detour and found myself on a small clear plateau with a tiny chapel and a building that I presumed must once have been the house of the resident priest. Both were now abandoned, and from their state of disrepair had been unoccupied for some time. I tried to get into the chapel, but found it boarded up, so I walked around the house and discovered the remains of a pool, dry and cracked, but obviously designed to catch the water from what must have been a very cool and welcome spring that now, in the middle of a dry summer, only just oozed out of a cleft in the cliff face. I sat on the stone bench beside it and watched the hills on the other side of the gorge pick up the light and warmth of the day. And it was while I rested there, content just to absorb it all, that I first became aware of a vague disquiet. At first I couldn't pin it down, but then as I looked from the spring, to the chapel, to the house, it gradually dawned on me that the chapel tower had no bell, and that the stairs to the rooms above the stable ran up the outside wall of the building. I was looking at something very like the scene my friend had described so vividly eighteen hours previously!

Then I remembered her alarm and began to feel uncomfortable. I stood and went over to the broad wood stairway on the house and thought about climbing to the living quarters. The treads were old and worn, but they seemed to be sound; yet I felt an extraordinary reluctance to climb them. I was afraid, and didn't know why, and this eventually made me angry enough to suppress the fear and start up the stairs. I had climbed perhaps halfway up the flight when I noticed that some of the dark weathered risers on the stairs were damp and shiny. I reached out to touch one and found it tacky; my fingers came away covered with a reddish smear that seemed vaguely familiar. I couldn't place it until I lifted my hand and smelled the unmistakable, chilling, slightly metallic taint of fresh blood. Then I panicked and ran. Nothing in the world could have persuaded me to go on up that stair. It was difficult enough even to turn my back on it; and I didn't stop until I was miles away down the Pleistos, well onto the Sacred Plain.

I met up with my friend that evening in Itea and she told me what else she had seen the previous day. While she had been describing the house—and it certainly was exactly as I had found it—a man in the black garb of a *pappas*, a Greek Orthodox priest, had come from the direction of the chapel holding in his arms the hideously disfigured body of a woman, and carried her up the stairs, trailing blood behind him.

Later we discovered that directly below the ravine with the chapel, on the northern flank of the river gorge, lies the pit of the Sybaris. In Greek fable this was a terrible monster which lay concealed in a cavern, emerging periodically to devastate the surrounding countryside, until it was destroyed by a heroic youth called Eurybatos, who hurled it into the ravine. Where it struck the earth, a spring gushed forth. And very much later, on a subsequent visit, I learned from local people that a shepherdess, the wife of a *pappas*, was found slaughtered near the spring, punished, they said, for having taken her flock to pasture there on a Sunday. Today they still know the spring and the whole ravine, which none of them care much for visiting alone, as the Pappadia.

I make no apology for including this personal story, not because I believe it proves anything, but because it illustrates in a dramatic way nearly all the problems of psychic research. The most vivid experiences of unknown things don't happen under laboratory conditions. They are usually spontaneous and unsolicited. The evidence for them is strictly anecdotal, and even the observations and all memories of them are highly colored by emotion and by the personalities involved. At this remove, I cannot even be certain that the blood was there at all. I believe it was, but all we are really left with is the similarity between my physical experience and my friend's visionary one.

Her detailed description of a place to which she had never been is possibly a good example of what can be achieved by out-of-body travel; but it also included a vision of an event that happened so long ago that it is part of local folklore. This makes it a retrocognitive experience, or perhaps even a telepathic awareness of something much in the minds of the people around us in the village. She might have practiced remote viewing of a real place, and seen there the ghostly presence of the priest, or stayed right where she was with me in the sanctuary and become possessed by the spirit of the widowed *pappas* still mourning his murdered wife. She also saw a scene which I was to experience myself in the flesh on the following day, which gives her vision precognitive or clairvoyant overtones. All of which makes it an out-of-body-visionary-telepathic-retrocognitive-precognitive-clairvoyant-apparitional-possessional experience. Unless of course you wish to trump all such notions with the single assumption that my friend must have been a reincarnation of the priest himself.

I see no point in applying any of the usual labels. The taxonomy of unusual events has helped parapsychologists to create some kind of order among the confusion characteristic of the psychic world, and may even have made it possible to design productive scientific approaches to it. But the results, even by the most sympathetic assessment, are meager, and the phenomena remain elusive. I suspect that the rigid categorization of events, and the subsequent painstaking search for them one at a time, has in fact obscured the most important feature of things supernatural—which is that they have holistic properties and can, if seen in their entirety, even be shown to have biological and evolutionary significance.

The French philosopher Henri Bergson, whose ideas are due for revival, suggested at the turn of the century that part of the vertebrate brain developed as a filter, the main function of which was to act as a screen, protecting conscious awareness from irrelevant external stimuli.[48] A few years later, in a presidential address to the Society for Psychical Research in London, he included telepathic signals from other people, and clairvoyant information from inanimate objects, in his list of stimuli which he considered to be biologically irrelevant.[49] Half a century passed before the Bergsonian filter was actually discovered by two physiologists at Northwestern University, who found that electrical stimulation of a cat's brain stem could arouse it from sleep as peacefully as a pat on the head.[387] They called this area the reticular activating system and decided that it acted as a sort of sentinel, capable of spraying the thinking area, the cortex, with general alarm or arousal signals. We now know that this very old part of the brain—in man it is about the size of the little finger—has a complex two-way interaction with the higher, more advanced areas. It is concerned with encouraging or inhibiting the flow of sensory stimuli from both outside and inside the body, and it regulates arousal, vigilance, sleep, and dreaming.[358] It is everything Bergson wanted it to be; a first line of defense against stimuli such as random telepathy, which most of the time has little or no meaning for us, and perhaps even against wasteful motor impulses of the kind which might be responsible for psychokinetic responses.

It begins to look as though the few perfect hits which have been scored in laboratory tests for telepathy, and which have been much vaunted in the parapsychological literature, may in fact be no more than mistakes. They occur due to flaws in the screen and have no organic significance at all. This is hardly surprising when one considers the boring, repetitive, mechanical tests with cards, dice, and electronic equipment which form the body of the experimental effort. Most of the approaches have no relevance to a living organism and in no way reflect any of its normal needs. This bias has been intentional, introduced it seems because it was assumed that no other scientists would accept demonstrations of the occurrence of strange phenomena unless they were totally isolated from the sensory and motor processes of the subjects—which is rather like asking singers at an operatic audition to perform with their mouths shut.

So laboratory conditions, as they have so far been prescribed, are unlikely to reveal anything more than random, capricious, trivial events which make no biological sense. Jan Ehrenwald suggests that the misses in such tests may in fact be more significant than the hits.[151] He chastises parapsychologists for being apologetic about the poor quality of much of their evidential material, and for straining to demonstrate the statistical significance of only those responses which demonstrate perfect correspondence with the target material. And he points out that in telepathic attempts to reproduce drawings, many of the receivers produce designs which are distorted in exactly those ways one would expect them to be if they had in fact been processed by the free-wheeling, nonlinear, unconscious areas of the brain. He is saying, in effect, that there is indisputable proof of telepathic communication in just those parts of the results which most experimenters discard.

Patients with brain damage of the sort which produces an imbalance between the cerebral hemispheres suffer from agnosias, which make it difficult or impossible for them to process knowledge in the normal way. Dislocation of the left brain produces dyslexia and agraphia, an inability to read and write; while disruption of the right brain results in visual disturbances. Those who have pictorial problems end up producing drawings which are vague and scrambled, and look exactly like the scribbles made by the poor confused subjects at the receiving end of attempts to send pictorial information telepathically under laboratory conditions. All they have to go on are the few scraps of information which occasionally manage to slip through the screens, when the filter is momentarily distracted by something else.

It seems clear that tightly controlled, artificial tests of the classic parapsychological kind have little or no biological relevance, even if they do sometimes succeed. The question which now needs to be asked is why they ever succeed at all. Why should our brains have the capacity to pick up any telepathic information? What good is it that some of us can sometimes bend spoons and door keys? For an answer one has to leave the laboratory altogether, and go into the forbidden field of emotion and spontaneous experience, looking for meaning in the random, unrepeatable, things which sometimes happen to people and which might prove to be what psychiatrists call "need-dependent," which, in biological terms, means having survival value.

Isolated events are difficult to analyze, but series of similar happenings begin to make some kind of sense. Spontaneous out-of-body states, for a start, seem to be directly connected to stress. Those who now do it consciously all report that their first such experience took place in a life-threatening situation. John Lilly says, "I was seven years old and I was having my tonsils removed under ether. I was extremely frightened as I went under the ether and I immediately found myself in a place with two angels."[347] Later he experimented with this ability under the influence of drugs, and was eventually able to slip out-of-body at will. The psychic Ingo Swann, who has apparently accomplished deliberate remote viewing in the laboratory while attached to physiological apparatus, had his first dissociative experience as a result of a childhood trauma.[521] The friend with whom I traveled in Greece had her first experience during a motor accident, apparently leaving her body even as the car in which she traveled was somersaulting at speed off a highway. She can now, it seems, do this whenever she wants to, but finds that when it does happen spontaneously, as it did at Delphi, there are usually good and urgent reasons for it. Neither of us has been able to discover why our experiences on that occasion seemed to coincide, or at least to reinforce each other. But there may have been some good life-supporting reason, such as the presence of a homicidal maniac with a cut foot in the abandoned house. I don't believe it was that simple, but there is a fair amount of anecdotal evidence to show that premonitions of disaster are often acquired during an out-of-body experience.

Rex Stanford of St. John's University in New York suggests that far more information gets through the brain screens than we are ever aware of, and that all complex organisms profit constantly by information that has been acquired paranormally.[504] He gives the example of a retired army colonel who unconsciously got off the New York subway at the wrong station, realized his mistake when he reached the exit, and was about to return again to the trains when he bumped right into the very people he was intending to visit. On a less dramatic level, there may be dozens of occasions every day when we receive an input and act on it, perhaps even psychokinetically, without being consciously aware of either the information or the related action. In an attempt to catch what he calls "psi mediated instrumental responses" in action in everyday life, Stanford devised a situation in which subjects were given a dull, repetitive task to do in one room while a random-number generator was operated without their knowledge in an adjoining room. The situation was designed so that the subjects would automatically be relieved of their boring duties as soon as the generator scored seven correct hits, matching its display against a predetermined sequence of numbers, in any run of ten trials.[505] The theory was that there might be information in an organism's environment which would be advantageous to it, but which it could not know about in the normal way, and that the remote existence of this information could be just the sort of pressure needed to encourage the use of a paranormal process. The work goes on, but to date he has found eight subjects who were apparently able to influence the apparatus strongly enough to escape within forty-five minutes. Left to its own devices, the generator alone doesn't score seven out of ten more often than once every two or three days.

This is a welcome attempt to make laboratory testing meaningful in some organic way, but it suffers from the same drawbacks as much of the earlier parapsychological research. Almost all experimental designs seem to have been blind to the real implications of the faculties they set out to explore. Implicit in each are the shortsighted assumptions that only the person who has been set aside as the subject of the experiment will use the talent under investigation, that only this one ability will be manifest, and that it will be demonstrated only in the prescribed way. There is no evidence that such restrictive standards have ever been met. But there are masses of evidence to show that in virtually every study ever made, the experimenter is more involved, more highly motivated, and more likely to influence the outcome of the test than the subject.[315] Rhea White has recently made an extensive survey of the possibility of such a bias in parapsychological testing, and concludes that absolutely no limits can be set to the degree of influence a tester has on his own results.[584] In many cases this is total.

The analytical approach of much of this work is still conditioned by the causal concepts of the old prequantum physics, in which the world is thought to be made up of points, each with a separate existence. But this doesn't work with telepathy, which, though still mysterious, seems to be a much more inclusive phenomenon, knitting things together in totally non-Newtonian ways. There are no simple causal chains, no logical connections. And any attempt to find them inevitably comes up empty-handed, open-ended, riddled with disturbing ambiguities. What little we have discovered is that conscious and unconscious processes play a very large part in everything that happens, perhaps even determining the nature of events that in ordinary terms have already happened.[469] There is no such thing as a random number; not if there are awarenesses involved.

It seems that the best we can manage by way of an experimental approach is a sort of updated version of the Bunyoro *segeto* stick. We have to have a scapegoat. Almost anything will do, and it doesn't seem to matter too much even if we consciously know that we have set it up as a prop for the duration of the investigation.

When table-turning was all the rage, several attempts were made to provide scientific, nonspiritual explanations for the apparent levitation. The most prestigious of these came from Michael Faraday himself, then in his sixties, who tried "turning the tables on the table turners" with a letter to *The Times* in which he attributed all the movement to "quasi-involuntary muscular action."[279] He apparently hoped by this interaction to kill off the craze with ridicule, but he probably gave it even greater impetus, because his theory was so obviously inadequate to those involved, who already knew that tables leaped around even when they were too heavy to be shifted by the combined conscious muscular action of all the sitters. A new study a century after Faraday's abortive attempt shows that it isn't that easily explained away. There actually is something strange going on.

Kenneth Batcheldor, head of the psychology clinic at a hospital near London, has found that movements and rappings take place despite the most rigorous controls.[35] In fact, he suggests that the only way to guarantee that nothing at all will happen is to pretend to those involved that the apparatus will not only prevent all movement, but will also pinpoint anyone responsible for producing previous phenomena. He attributes this psychological stopper to "ownership resistance" and points out that nobody can make a Ouija board work on his own. It takes at least two to play, because then it is possible for each participant, even if he is himself unconsciously responsible, to blame it on the other. There has to be a scapegoat. In a classical séance, the sitters blame it all on one of their number who is designated as the medium; and that person, depending on his particular bias, passes responsibility on to spirits, ghosts, gods, or demons as the case may be. For no matter how real our interest in the phenomena may be, it seems we all have a latent fear of actually becoming involved, of being responsible. One of the best ways around this, Batcheldor found, was a technique he calls "artifact induction." Without the knowledge of the other members in the group, he deliberately and secretly moves the table himself by ordinary muscular force. This gets the others excited, ready to believe in what is happening, and soon, "The phenomena, at first superimposed on artifacts, gradually strengthen until they are clearly paranormal."[35] In the end, he and his group were able to levitate heavy tables and even a piano without touching them at all.

As the group grew more adept, other unintended phenomena began to intrude. Small foreign objects began to fall out of nowhere like classical "apports." One stone four inches in diameter was sent to a London museum for analysis and never heard of again. Batcheldor got a bewildered note from museum officials saying the rock had disappeared.[83]

Inspired by these successes, other groups were established in England under the guidance of an engineer, Colin Brookes-Smith.[81] He was primarily interested in taking quantitative measurements with height gauges, strain gauges, dynamometers, and modulation amplifiers, and found that he could get readings on all his instruments, but that the results were always best if the scrutiny was covert and if he followed Batcheldor's advice and simulated the setting of a Victorian séance, which was often a great social occasion, with food and drink, gaiety and song. It seems that table-turning phenomena are produced by psychological skills, and need to be encouraged in some indirect way, partly by suspension of disbelief in a relaxed atmosphere devoid of the obvious trappings of science, and partly by absolving all those present of responsibility or blame, by suggesting that it is the table and not they that does such things. Working on this basis, it seems that almost anyone can, with patience and the right approach, wreak havoc on the laws of physics with any old piece of furniture, or, using rod or pendulum as a scapegoat, learn to practice the equally mysterious art of dowsing.[215]

As the word gets around, new groups of perfectly ordinary people, without any psychic pretensions, are producing paranormal phenomena simply by pretending that the rules of science don't exist. I have sat with a group in Toronto who blame what they do on an imaginary ghost they call Philip.[418] By consciously attempting to behave like children, by singing silly songs and regressing to a point where their communal thinking once more takes on magical qualities, they regularly, even on live television, produce levitation and rapping sounds which they, tongue in cheek, insist are Philip's responses to the bawdy conversations they hold with him. Not the least astonishing thing about the effects is that the sounds have been proved to be genuinely paranormal, having peculiar acoustic qualities and lasting only 0.16 second, which is about a third as long as the sound you make by rapping a table with your knees or knuckles.[586] A few miles away, another group communicates with Santa Claus; and a third committee of conscious poltergeists have adopted as their whipping boy none other than Dickens' Artful Dodger.[379]

It is gradually dawning on us that we all have this potential. Paranormal phenomena are part of the normal repertoire of human behavior and can be produced on demand, at will, given the right circumstances. Anyone can learn. It even seems likely that genetic factors may play a much larger part than is generally assumed, giving each individual a greater or lesser ability to use the Bergsonian filter selectively. When we are young, the filter tends to be less stable, it too has to be trained, and there is an increased chance that erratic, uncontrolled, and potentially destructive leakages can occur. It is certain that all poltergeist phenomena center on the person of an individual, usually a child, and most often an adolescent girl, undergoing a period of particularly difficult emotional adjustment. Things began this way for Matthew Manning in Cambridge, but he has shown that even these manifestations can be trained and transformed—sublimated in his case into artistic and creative activity of a high order.[359]

The chances are that all mediumistic phenomena, all spiritualist goings-on, have their basis in filter flaws, in leakages to and from unconscious areas which we all seem to have in common. Most, if not all, paranormal events need people. They are closely related to human personalities, to attitudes, to states of consciousness; to energies or images released from, or inspired by, the unconscious; to powerful imaginations and strong wills. There is no need to conjure up spirits, no need to seek the services of a medium. There is little real evidence of ghostly intelligences at work. Kenneth Batcheldor, who has done more than any other to lay these particular ghosts, says,

"I think it possible that 'mediumship' is no more the prerogative of freak or abnormal personalities than is hypnotic behaviour, which was originally thought to be confined to hysterics . . . The rarity of a behaviour often suggests rare personality, but later it turns out that it is the conditions for the occurrence which are unusual."[35] He has defined these conditions very carefully in a series of practical hints for anyone interested in setting up a small group to explore their own potentials.[36]

There have been any number of valiant attempts to make scientific sense of what happens in apparently psychic situations. Electromagnetic, nuclear, gravitational, acoustical, and chemical explanations have all been evoked. Many intellectual mountains have labored mightily at the task, but we seem to have ended up with nothing more than a collection of lame mice. Physicists have toyed with the notion of parallel universes and quantum mechanical tunnels in hyperspace, with backward flows in time and ghost particles of negative mass. With the aid of mathematics, they can build elaborate theoretical structures of great beauty and power, which are internally totally consistent and make it possible to discard all commonsense notions of reality. But these provide little satisfaction for those not privy to the secret language of numbers, and seem to have been singularly ineffective in bringing the phenomenon under control, even to the extent of being able to predict what might happen next. Philosophers have produced challenging hypotheses to bridge the logical gap between mind and matter, running the gauntlet of mentalism, materialism, parallelism, interaction, emergence, and downward causation, only to end up concluding that it is "improbable that the problem will ever be solved."[438]

Biologists, with a few honorable exceptions like Alister Hardy, have largely ignored the problem altogether; but I believe that a liberal vitalist attitude, broadly based on all the life sciences, offers the only chance we are likely to have, not necessarily of an explanation, but at least of a limited understanding of what is going on.

Arthur Koestler finds meaning in the basic polarity of nature itself, in alternation between the two processes of differentiation and integration. He points out that in a growing embryo, successive cells become specialized in different ways, branching out into diversified tissues, but that these then come together again in integrated organs. These in turn come to be connected into a living individual who is more than a mere assembly of isolated bits. Organisms, says Koestler, "are multi-levelled, hierarchically organised systems of sub-wholes containing sub-wholes of a lower order, like Chinese boxes. These sub-holes or 'holons' . . . are Janus-faced entities which display both the independent properties of wholes and the dependent properties of parts."[321] So the complex is maintained in a state of delicate equilibrium between the self-assertive tendencies of the parts and the integrative tendency of the whole.

Such polarity is basic to all processes in the universe, which is balanced between centrifugal forces throwing everything to the cosmic winds, and cohesive forces such as gravity, holding parts down in definite, albeit dynamic, relation to one another. In biological terms, the equivalent of gravity is the life field, *élan vital*, soul substance, call it what you will, which brings the parts into meaningful relationship with one another. But we still haven't the faintest idea about what it is or how it actually works.

High-speed film of integrated flocks of flying birds shows that as many as fifty thousand individuals can turn in synchrony in less than one-seventieth of a second. If the birds were playing follow-the-leader, either watching or waiting for a signal, wave effects would show inside the flock; but they don't. Neither does it seem likely that a leader sends out signals which are picked up by the followers and translated into immediate action, because shifts in leadership are constant and not at all clearly defined. The best current theory is that birds have complex mechanisms in their feathers which act as radioactive and microwave receivers and transmitters, linking the active flock into one "superindividual," and that it is lateral differentials in the way the electromagnetic field acts on each bird, that weakens one wing or the other momentarily, sending the flock dipping and wheeling in a patternless way.[351] There is still little evidence that anything like such a mechanism exists, but the theory contains an important idea, which is that collections of cells, or groups of organisms, can fit together into a functional whole, which responds unconsciously and collectively to certain stimuli, and demonstrates properties unknown to any of the separate components. The whole is greater than, and very different from, the sum of its parts.

This is by no means a new idea, but I think it is one worth reviving and giving perhaps a slightly different emphasis. Earlier this century, in the wake of Bergson's theory of creative evolution, a form of vitalism described as "emergent evolution" was put forward. Essentially this said that when two or more simple entities come together, they may add up in unexpected ways. Most biologists today, affected as we all must be by the discoveries of molecular biology and quantum physics, believe it is the arrangement of atoms that counts, and that function is determined at the molecular level. But any field naturalist must know that this is not the whole answer. There are powerful ways in which the total form clearly exhibits functional qualities of its own. It is impossible to predict the nature of the hive from the behavior of even a large number of solitary bees.

There are, as I have tried to show, a number of creative thresholds evident in the process of evolution, beginning all the way back with the first accumulation of significant quantities of interstellar dust. Each time a gathering of organelles, or organs, or organisms, reaches a certain critical mass, it takes on novel properties. It transcends itself. It enjoys the biological equivalent of what Abraham Maslow calls a "peak experience."[368] In human terms this means a sense of openness and freedom, a feeling of belonging, just for a moment, to something bigger. For a reason no one quite understands, this is in itself rewarding. It might be initiated by the sound of music, by the sun on a hummingbird's wing, by the scent of hibiscus on a summer evening. Or it may be sparked internally by drugs, devotion, or hypnotic rapport. Whatever the stimulus, the result is the same—a craving to do it again, to transcend personal boundaries and surrender to the "sympathy of all things."

Perhaps all living entities react in the same way. Maybe the directional force, the creative element in evolution, boils down to this one simple genetic tendency: an inherent predisposition to react positively to the threshold. And because the DNA-controlled immune system is more concerned with differences, with picking out the "not-me" and rejecting it, and is anyway a relative newcomer, I suggest that the ability to respond to the "me" in others is primarily under the control of the contingent system, which is something all organisms have in common. Regardless of their genotype, a whale, a wasp, and a willow all have identical organelles. At the level of their mitochondria, they all speak the same language.

Our lack of scientific success in bringing paranormal phenomena to heel strongly suggests that these obey different laws, and can't be approached on a causal basis. So I am not going to set up the contingent system as a causal factor. I certainly don't think the answer is going to be that simple; but I do believe that the existence of an alternative adds significantly to Koestler's contention that we are in a state of essential tension between rival forces. This is the conflict that produces what I have called the Lifetide. And I emphasize again that the tidal metaphor is appropriate, because tides and waves are phenomena that have nothing directly to do with the water in which they become manifest. They are patterns passing by, and not even an atomic analysis of a test tube of wave substance will tell you anything about this nature. The tide, and all ideas about it or about anything else, exist in a world where space and time are not appropriate descriptions of the state of being. We will never be able to measure the weight, length, or duration of an idea, but we can measure the strength of our reaction for or against it. We can use it to learn about ourselves and others.

There is an important contradiction in the apparent ease with which anyone can learn to dowse or to levitate a table, and the rarity with which such phenomena occur spontaneously. This rarity may be only apparent in that we simply don't notice dozens of strange things which occur around us every day, but even taking that possibility into account, it nevertheless seems clear that our internal filter normally exerts a very powerful two-way control. This suggests that paranormal events have limited survival value and a narrow field of applicability, which leaves us in the awkward position of having to make major assessments on the basis of fragmentary evidence. What we can see is nothing like the tip of an iceberg, which does in fact offer an excellent idea of the composition of the rest, but merely the debris left behind by a series of unsuccessful experiments. We have to operate like plastic surgeons faced with the task of resurrecting a damaged face without a photograph of the original to refer to. Perhaps the best analogy is that of the psychiatrist with a severely disturbed patient, who has to search for health in the evidence of pathology. All he has to work with is what the patient tells him. All we have to work with is what seeps past the barrier, what little the unconscious, in unguarded moments, lets slip.

For several years now I have been collecting together all the available evidence for the activity of the collective unconscious, with the intention of measuring it on its own ground, of applying classical psychoanalytic techniques to what it actually says and does. I have been actively involved with groups working with or through mediums. I have collected millions of words produced in the form of automatic speech and writing. I have recorded hundreds of recollections and far memories of apparent previous lives produced under deep hypnosis. I have listened to hours of taped voices which seem to have been plucked out of thin air. I have interviewed those who believe they have lived before, and those who claim to have come from other planets. I have bulging files full of letters in several languages from people in various states of dissociation and delusion, written on both sides of ruled paper, heavily underlined and profusely illustrated in at least three colors. And I have begun the daunting task of trying to make sense of it all.

A very small fraction of the material is logical, precise, and to the point. The "teachings" of Seth, the poetry of Patience Worth, and the beautiful "Book of Matan" fall into this category.[135] So too do the writings of Joan Grant, the Brazilian Chico Xavier, and a few other gifted automatists like Edgar Cayce.[433] Some of the communications make sense on their own terms; they are internally self-consistent, such as those obtained by the Ossining group with Phyllis Schlemmer.[262] And a minority of others are capable of sympathetic interpretation in the manner of William Butler Yeats' reworking of his wife Georgie's trance material.[598] But the vast majority of anything I have ever seen or heard is rambling and incoherent.

After months of analysis, I am able to draw these few conclusions: The content of the material is almost unremittingly banal, consisting mostly of pious sentiments, couched in woolly language, entirely consistent with that used by those involved in providing the information. There are the expected examples of cultural bias, and an interesting tendency for those concerned to get what they want to hear. In the last ten years, for instance, the emphasis has shifted noticeably from the "next world" to UFOs and Atlantis; and then moved again more recently to a more general concern with higher consciousness and good ecology. There is in several cases evidence of a fair knowledge of Gnostic philosophy and of esoteric teachings, in particular of the Jewish Qabala. And in a few examples, obvious familiarity with the work of J. R. R. Tolkien. Throughout the material, Jesus of Nazareth tends to be referred to with great respect, usually as "The Christ." In short, there is nothing there that would not be normally available to those involved, or to someone near them, or which reflects beliefs markedly different from those they themselves hold.

As far as style is concerned, there is with very few exceptions a tendency toward circumlocution. Even when messages are being spelled out laboriously on an Ouija board, they tend to be vague and wordy, often evasive. To give a typical example, a mutual friend of my brother Andrew and I received this, in automatic writing. "An important communication wings to you, but not for you, from the land of the lily bearing that name."[18] The information was correct. I had written to this person urgently from Bermuda, asking him to pass the letter on to Andrew, whose address I had lost. So why not say so? Freud perhaps put his finger right on the answer. If telepathy was a fact, he concluded later in his life, then the laws of unconscious mental life could be taken for granted as applying to data telepathically perceived. "Distortion of perception," said Freud, "is one of the characteristics of mental functioning dominated by unconscious needs."[136] And one of the most frequent distortions of the unconscious is toward timelessness and vagueness.

I found that after a while I lost track of which productions I was reading at any one time, because there is a basic similarity to them all. If you cut out local evidential detail, answers to specific questions, and obvious idiosyncrasies produced by the particular medium involved, there is a sameness in the tone, the word structure, the feeling, and the delivery of almost all the material. It has a dreamlike quality, and my feeling is that the vast majority of all the evidence I am looking at is a series produced by one prodigious dreamer. The manifest content of the dreams is variable, according to circumstance and environment, but it seems to me that in the latent content there are a small number of recurrent themes.

Jung believed that all paranormal phenomena were probably manifestations of the unconscious mind, whose deepest strata extend into the collective unconscious. He suggested that the decisive factors in the collective area were archetypes, which were sort of distilled racial memories, usually represented in elusive symbolic form and providing the common basis for all man's mythologies. He also thought that the archetypes provided a kind of genetic control in that they governed our behavior, surfacing into consciousness at times of stress with extraordinary effects; but more often working underground, molding our characters at an unconscious level by providing three major ingredients which he called the Shadow, the Anima/Animus, and the Self.[299] These, in very simple terms, represent our weakness, our sexuality, and our potential strength. I find elements of all these classical archetypes in the mediumistic material, but feel certain that they apply in every case to the character of the medium or to that of an associate or sitter. Psychiatrists are now well aware of this fact and several make good use of telepathic and clairvoyant material as convenient short cuts in analysis.[415] But, over and above these personal intrusions, there are also a number of other motifs which I cannot dismiss in this way, and which I believe have a basis which may be more organic. As I can find nothing in individual experience to account for the observed similarities, and no evidence that they are embodied in any way in genetic inheritance, I am assuming that they have some other more fundamental common origin.

I intend at some future date to publish a complete and detailed analysis, but it is relevant and important here to outline what appears to be four persistent themes which, because they seem to differ from and antedate the traditional archetypes of analytical psychology, I choose to call prototypes.

Prototype One is associated with space and air and stellar motifs. It deals with sensations of flying, with feelings of climbing, traveling, crossing, ascending, going on a journey. Personality is minimal and if it intrudes at all, gives rise to feelings of insignificance, of being small in the midst of greatness, of invisibility. If colors are included, they tend to be shades of blue; and if direction is apparent it is usually eastward, toward the sun. I suggest that awareness of this prototype is responsible for all out-of-body experiences, which, significantly, are often referred to as astral traveling. I believe that it can, when coupled with telepathic sensitivity, and the capacity of the unconscious to organize information in a personal way, account for all remote viewing and apparent personal knowledge of distant times and places. It is possibly behind the worldwide belief in, and lasting concern with, the idea of reincarnation. I conclude that it originates in the earliest stirrings of organized interstellar dust, that it is in effect a primitive biological memory of cosmic beginnings. And I have, for these reasons, called this first prototype the Seed.

Prototype Two is more down-to-earth. It is associated with darkness and soil and spiral motifs. It deals with sensations of falling, with images of depth, caves, descents into the abyss. Ghosts and shadows flicker on the edge of awareness in flashes of primary reds and greens, but are not necessarily fearsome, just unknown. If direction is indicated, it is northward into the night. I suggest that experience of this prototype is what ultimately produces the archetypal Shadow and all our unreasoning, emotional, reaction to the darker side of things; but I suspect that it is also what gives us our roots on earth. There are strong connections with Jung's Earth Mother archetype and perhaps a basis for man's recognition of, and fear of, the power of the feminine. I conclude that this prototype originates in the organization of early organic molecules by earth templates, that it is a memorial to our crystal ancestors in terrestrial clay. I call it the Soil.

Prototype Three is the least clearly defined. It is associated with heat and fire, with the creative and destructive power of the sun. It often takes the form of a phoenix, rising from the flame. It multiplies until it is peopled with teeming hordes, with monsters and giants. It deals in emotional terms both with pursuit and hiding in the manner of nightmares; and with escape and security in the company of family and friends. The predominant motif is a cross or a gate. The colors are golden and for those in the northern hemisphere they always lie in the south. My feelings about this prototype are vague. I think that it can probably be further subdivided and represents, as it stands, little more than a convenient hold-all for the confused biological memories of the whole sequence of events that add up to the evolution of life with all its complexities. It includes the Jungian archetypes of Earth Mother and Wise Old Man in all their variants, as well as the Animus and Anima. And, purely as an interim measure, I suggest that this third prototype be called the Flower.

Prototype Four is very clearly delineated. It is lunar, circular, acausal, ethereal. There are similarities with the Seed, in that it deals with openness, but the sensation is one of floating rather than flying. It is located in water rather than air, the feelings are oceanic, and personality is very much to the fore. Intellect predominates, combining with earlier intuitions to provide the blinding white light of understanding. This is the realm of pure mind, of psyche in the sympathy of all things. I suggest that awareness of this area is behind all peak experience and that it is not so much a memory as a premonition, a sense of what might be. It should perhaps be called a metatype rather than a prototype, and I choose for obvious reasons to refer to it as the Fruit.

Any classification of this kind is necessarily artificial. Most dynamic processes occur in a continuum with no sharp distinction between one area and the rest, but I believe that there have been in the course of evolution of life on this planet a number of quantum leaps which make my categories little more than mere administrative conveniences. There is a directional flow from Seed to Soil to Flower to Fruit, but there is also sufficient qualitative distinction between each stage to render it conspicuous enough to stand on its own, making a separate and recognizable impression. I suggest that there are moments, in the embryological development and later integration of every human personality, when we become aware of each of these historic stages and are influenced in many ways by them. And that it is in times of conflict, when integration is most difficult, when we are tugged this and that way by the conflicting interests of genetic and contingent systems, when the equilibrium between holons and the whole is disturbed, that this influence becomes most apparent.

I suggest, in brief, that we are motivated by a great deal more than genetic inheritance and environmental influence; that nature and nurture are not enough to account for what we are. We have to make room in the argument for a third participant with a new and revealing viewpoint. We need to begin to think seriously about the natural history of the supernatural.

# CHAPTER TWELVE

# PROVOCATION: Inventing the Future

Take the egg of a frog. A fresh one looks like a ball of jelly with a small dark spot. In this nucleus are half the normal frog number of chromosomes, waiting to join up with a matching set from the first convenient sperm. But it is important to note that fertilization is necessary only for fertility, not for making a new frog. The maternal chromosomes on their own contain all the relevant instructions for hopping, croaking, and catching flies. Even without male participation they can produce an individual capable of doing these things. It will be sterile, but otherwise identical to any other adult of its kind. And all that is necessary to start the train of embryonic growth rolling in this direction is a little prod, a hint from someone else. Normally the starting signal is provided by another frog, but even a camel can do it. It is enough just to prick the egg gently with a fiber from a camel-hair brush. That's all, and there you have it, instant tadpole.

An unfertilized egg is like a complicated question. In its genetic code is all the grammar necessary for posing the query, but the answer has to come from outside. The reply can be very simple—most of the best retorts usually are—but without it nothing happens. None of the potential is realized. The question the frog's egg is asking is "How can I, a radially symmetrical structure, set about fulfilling my destiny and become a bilaterally symmetrical, amphibious animal?" The correct answer is normally provided by a frog sperm penetrating the egg at a point near its equator; saying, in effect, "I suggest you start with a meridional division along a line passing through this point."

The fact that the sperm also carries genetic information is purely incidental. The critical stimulus is the one provided by its actual entry into the egg. In doing so it establishes one point on the surface as distinct from all others. It recognizes difference. It doesn't much matter what is responsible for making this distinction. The notion can be conveyed equally well by a camel hair, by the tip of a needle, or even by a sharp gust of wind. It is the idea itself which is seminal. Conception, in a direct sense, means nothing more than the introduction of a concept. In this case, the notion of difference.

But what is difference? It is actually a very peculiar and obscure concept. It is certainly not a thing or an event. The paper on which I write these words is different from the paper on which you read them. For a start they owe their origin to different trees that grew in different parts of the world; but as soon as we start to locate differences spatially in this way, we run into problems. The difference between two sheets of paper is not in the papers themselves, and it is obviously not in the space that separates them, nor in the time between them. A difference is an abstract matter; an idea. It belongs in a world without space or time; it enjoys an independent reality, but it nevertheless has the power to interact with the world of substance.

In the Northern Territory of Australia, the aboriginal Tiwi are adept in the art of pointing bones. They use the femur of a large lizard or the humerus of a pelican, sharpened to a point at one end, and attached at the other to a length of human hair. In matters of revenge, such a bone is pointed in an elaborate ceremony at the mental image of an enemy, like an ice pick brutally applied to the delicate membrane of a unfertilized egg. To reinforce the effect of this lethal ritual, care is often taken to let the intended victim know about it. He is told of an intervention in the equilibrium of things; he is made aware of difference. And it makes him feel different. The mere knowledge that a spell has been cast is often enough to produce disruption and death by enchantment, a process in which modern medicine usually finds itself powerless to intervene. But there are also several cases on record of sudden death befalling unwitting victims who had yet to hear, at least consciously, of the bone aimed at them from a distance.[184] An idea can, it seems, have the power of life or death. It is a biological unit in its own right.

We have every reason now to doubt René Descartes' notion of a fixed "out there" and a separate "in here." There can be no hard-and-fast boundaries drawn between organisms and their environment, when we know that psychokinetic effects occur, and that a few people can even "think" images into chemical reality on photographic film, or into magnetic existence on a tape recording.[157] I have experienced both phenomena at firsthand under circumstances where I feel certain no other explanation will do. Given these facts, we are compelled to reexamine all definitions of mind, which see it only as a nebulous entity at the end of a one-way street of sensory traffic. When we learned that the brain exercised its own selective editorial policy, monitoring input and ordering priorities to suit particular needs, we had to adjust our assessment of perception. In the same way, it now becomes necessary for a comprehensive reappraisal of the role of conception in structuring reality.

Quantum physics already includes consciousness as an essential hidden variable in its equations, but these purely mathematical accommodations are unlikely to make much difference to the way in which most of us deal with reality on an everyday basis. And yet if it is true that we are all so intimately involved in the creation of reality, then there must be some sort of unconscious feedback to us from our constructs, which determine at least in part how we continue to think about the environment, and what kind of structures we maintain there. If we evolve, then reality must move along with us. Is there in fact any evidence that this is happening? I believe there might be.

Lord Ritchie-Calder defined science as "the everlasting interrogation of Nature by Man."[95] This sounds grand, but it doesn't tell the whole truth. The process certainly began with the first question "Why?" But then came the method. The first practitioner noticed something, and became an observer. He looked again, to make sure, and became a searcher. He drew inferences from the facts, and became a theorist. And then he tested his theories against the facts, and became a researcher.

The last step was the critical one, because it led not only to an assembly of ideas and to the development of natural philosophy, but to an exchange of ideas with others holding the same or rival theories. It gave rise to the dialectic, the politics of science.

Peter Laurie points out that scientific truth today, "which is supposed to be completely independent of time, place or person, is actually based entirely on political or social evidence"[337] and not on observation or fact.

For example, I believe in the electron as a basic building block of matter. But on what evidence do I base this belief? I don't have, and even if I did wouldn't know how to work, the apparatus necessary to demonstrate its existence. Instead I take the word of the Wykeham Professor of Physics in the University of Oxford, and I ignore the opinion of the Professor of Cosmic Knowledge in the University of Light, who insists that electrons are actually tiny transformations of the souls of the departed. Experiments about electrons boil down in the end to experiments about professors.

And why do I choose to believe the one rather than the other? Simply because the Oxford opinion is one verified by an elaborate system of rival researchers who duplicate, largely with the hope of discrediting, each other's work. A system of commenting and criticizing, weighing, assessing, and refereeing in which experts sit in judgment on the facts, ultimately reaching a consensus, selecting what is true and rejecting the remainder as false. But, in the final analysis, this is a political process, not a scientific one.

Somewhere between the question and the answer there is room for opinion. The answers are only approximate, and Knowledge, it seems, is very largely a matter of belief. The rigid experimental protocols, the strict scientific procedures which were thought to hold individuality in abeyance, have probably always been contaminated. Total objectivity may be nothing but a myth.

If this is so—and few of those involved in any way with quantum physics now deny that the observer, simply by being there, influences the outcome of the experiment—then we face a dilemma.

If faith is part of the scientific process, lying clearly in the path between question and answer, then might it not also intrude between Nature and the question? Is it possible that Man, by his interrogation, changes Nature itself?

I believe he can, and does.

Consider one dynamic paradigm—man's changing view of the solar system. Hipparchus, the greatest of Greek astronomers, developed in the second century B.C. the first comprehensive scheme in which all the observed motions of the planets were accounted for by a complex of epicycles based on the implicit assumption that earth was the unmoving center of the universe. Why go to all this trouble with difficult mathematics when Aristarchus, a century earlier, had proposed a simpler solution in which it was assumed that the heavenly bodies revolved around the sun? For two reasons. First, the scheme of Hipparchus worked. Using it, he was able to predict the positions of all known planets at any given time. This was important for astrology and the timing of rituals. And second, it is very hard to think of the whole earth flying through space, unless you have been taught this description of reality as a child, and will believe anything.

Claudius Ptolemy, two centuries later, carried on the work

of Hipparchus and elaborated it further into a geocentric system that came to be called Ptolemaic and which lasted, and functioned very well, for fourteen centuries. It was only displaced in 1543 on publication by Copernicus of a mathematical system that could cope with the calculation of planetary orbits on a heliocentric basis, and come to terms with the notion of an earth that moved.

In 1977, a book by Robert Newton of Johns Hopkins University suggested that Ptolemy was a fraud.[407] Working backward from modern astronomical tables, this twentieth-century Newton found that Ptolemy's observations were so accurate that they could never have been made with the instruments and attitudes described in his great thirteen-volume work *The Almagest*. Robert Newton concludes that Ptolemy's work did "more damage to astronomy than any other work ever written, and astronomy would be better off if it had never existed." Ptolemy, he suggests, is "not the greatest astronomer of history, but something still more unusual; he is the most successful fraud in the history of science."

There is cause to quibble with Robert Newton about the deleterious effects of Ptolemy's geocentric theory. One of the basic tools of new physics is the correspondence principle, which suggests that every new theory, if it is to prove useful and secure knowledge without undue loss of past achievement, must offer a limiting transition to the old theory which it replaces. Or else, in terms of a modern metaphor, you end up tossing the baby out with the bath water.

But a more fundamental criticism of Robert Newton's strangely shrill attack on someone who died two millennia ago is that it could be based on a fallacious assumption: namely, that if he is right, Ptolemy must be wrong, and therefore dishonest. It does seem strange that the Greek astronomer, holding hypotheses which we see as invalid, and working with instruments which we consider inadequate, should nevertheless have arrived at what we believe to be the right answers. At least it appears odd unless we change our basic assumption and suggest instead that Robert Newton and Claudius Ptolemy both reached the same answers using different tools, because they used them on different models of reality. And that neither of these models is necessarily the "right" one, but that both bear some sort of intrinsic relationship to the true nature of things.

In other words, all facts are laden with theory, and Nature has a way of turning out to be partly what is, and partly what we want it to be.

But that's heresy, so let's return for a moment to our shifting paradigm. One hundred and fifty years after the death of Copernicus, his findings, and those of Galileo after him, were codified into the famous three laws of motion by the true Newton, Isaac, who, in *Principia Mathematica*, made an intuitive leap to the notion that the same laws applied to earthly and to heavenly bodies. With the aid of his law of universal gravitation, Newton developed an overall scheme of things which was so precise, so elegant in its construction, that, almost unaided, it ushered in the Age of Reason.

If anyone slowed the growth of astronomy it was Isaac Newton himself, who cast such a long shadow that many of his heirs thought he had left them nothing new to discover. Another century and a half passed before William Herschel in 1781 brought a breath of fresh air to the science by discovering a new planet, Uranus. In the following sixty years it became apparent that the new planet was not behaving precisely as Newton's laws demanded, and in 1856 John Adams in England and Urbain Leverrier in France, working without knowledge of each other and reasoning only with pen and paper, decided that the irregularities in the orbit must be due to the presence of another planet beyond Uranus. Both calculated where it should be visible and on the very first evening that a large enough telescope was turned to the prescribed spot, there it was, right on cue. And we now know it as Neptune.

This discovery is usually seen as a triumphant endorsement of Newton's laws, as further proof that all problems can be solved by careful observation and the skillful use of mathematics. But it seems to have been overlooked that both Adams and Leverrier were calculating partly on the basis of Bode's Law (a now disproved notion that the planets orbit at distances from the sun which can be predicted by a scale of proportions which rise by a constant increment in this way: 4 : 7 : 10 : 16 : 28 : 52 : 100 : 196 : 388 : etc.) If it is assumed that the asteroid belt is the remains of a planet that once lay between Mars and Jupiter at position 28, then all the inner seven planets fit the scale exactly. So it was generally assumed that, if there was an eighth planet, it would be found at position 388. But it is nowhere near there. Neptune is much farther away, and yet, on September 23, 1846, when Johann Galle aligned the Great Berlin reflector according to instructions, the planet was right in place, on demand.

Adams and Leverrier, using the wrong tool, making the wrong assumptions, came up quite independently with the right answer. Should we now begin looking for an international conspiracy to defraud? I doubt it. But I am very intrigued to know that, since the discovery of Neptune, it has been found that even those famous calculations didn't take all the discrepancies in the motion of Uranus into account.

Uranus still tends to wander off its predicted orbit by a fraction, so in the last years of the nineteenth century Percival Lowell turned the resources of his private observatory in Arizona over to a search for yet another member of the solar system. He called it Planet X. Precise calculations predicted where it ought to be found and a careful search was made, but the ninth planet didn't materialize until fourteen years after Lowell's death.

On March 13, 1930, Clyde Tombaugh—then a young unqualified assistant in the Lowell observatory—finished a year of painstaking picking through comparative photographs of the critical part of the sky. And there, moving almost imperceptibly across a field of four hundred thousand equally faint stars, was Pluto, god of the nether darkness. It is no accident that the first two letters of the new planet's name should be the initials of the man who decided where to look. And it seems fitting too that, following Tombaugh's discovery, it was revealed that Milton Humason of the Harvard Observatory had a few years previously taken a photograph of the precise location where Pluto should have been, but seen nothing there. This mysterious failure is officially attributed to the fact that Humason must have succeeded in obtaining the image of Pluto —after all, anyone can do it now—but that it fell right on a tiny flaw in his photographic plate.

I am well aware of the pattern of synchronicity in scientific discovery; of the frequency with which two or more researchers, apparently without collusion, simultaneously produce answers to questions that seemed insoluble for years. And how, once the barrier is broken, the solutions often seem so painfully simple it is difficult to understand why they weren't obvious to everyone right from the start. I am not necessarily suggesting, by presenting a brief history of our discovery of the solar system in this way, that the outer planets didn't exist until we began to look for them. But neither am I prepared to dismiss this possibility out of hand.

There is room for similar doubts at the other end of the cosmic scale. In fact the difficulties there are even greater, for despite quantum leaps in technology, it is still true that no one has ever seen an atom. Yet physicists confidently talk about its subdivision into hundreds of elementary particles—and this number is being added to at the rate of about two a month. In the Alice-in-Wonderland of the nucleus, a bewildering variety of short-lived entities—leptons, hadrons, and quarks, with whimsical properties such as "strangeness" and "charm"— come and go like the cast of thousands in a Hollywood spectacular. This embarrassment of riches has led some of the scientists themselves to raise the heretical question "Are these things actually part of Nature, or did we invent them?"[291]

Joseph Chilton Pearce points out that when the neutron was first disintegrated, the resulting particles, like planets which refuse to bow to the laws of Newton, were found to be behaving badly. Something had to give; and since it couldn't be the law of the conservation of momentum, on which too much else depended, another scapegoat had to be found.[427] It was the Italian physicist Enrico Fermi who finally came to the rescue. He postulated, in addition to the electron and proton, a third particle of no charge and no mass, which he called the "neutrino" or little neutron. This mysterious newcomer, without much of anything, was later decided to have a spiral orbit, and eventually, when enough concrete properties had been ascribed to it, it was finally "discovered." When evidence for its existence took on a sufficiently large number of the accepted aspects of reality, it materialized like Philip, the imaginary ghost.

Whenever a new particle turns up in a laboratory experiment, other researchers try their hands at looking for it too. More often than not, they find it, sometimes right where they looked before without success. Efforts are also made to double-check with Nature by trying to find the particle in natural cosmic rays; and sometimes it turns up there as well. But the doubts remain. All the methods of detection are man-made. In detecting, we may be creating that which we seek to find.[171] Once again it seems we lie between Nature and the question.

Cyril Hinshelwood, a Nobel laureate in physical chemistry, has suggested that a more appropriate name for the particles might be "manifestations."[255] That sounds right. In purely physical terms they have little reality. Some are light, some heavy, some stable, some ephemeral; they appear and disappear; they differ in mass and charge, in duration and intensity. Only in the poetry of mathematical symbolism is any kind of consistency apparent. Perhaps they exist only in consciousness, and can be made manifest only in retrospect, like the memory of a dream. Perhaps too, as in dream interpretation, we should be paying more attention to the image and wasting less time on the words.

William Blake said, "Man's mind is like a garden ready planted. This world is too poor to produce one seed." He attributed all to "divine genius," which may be giving us and our earth too little credit. But, as usual, he was right about the strength of our innate ideas, about our capacity for creative and original thinking, independent of mechanical information from the world. If an imaginative seed, nothing more than an idea, can be planted, even if it is contrary to all existing evidence, it tends to grow; and sooner or later it seems to be capable of producing its own confirmation. Data can eventually be found to reinforce its existence, to bolster the conviction of its reality. The desire for conviction produces its own data, its own manifestations, even its own visible physical demonstration.

Tibetan mystic lore contends that the whole universe is a mirage. A question often put by monastic philosophers to their pupils goes something like this: "A flag moves. What is it that moves? Is it the flag or the wind?" The answer, according to them, and in accord with modern neurophysiological and quantum mechanical thinking, is neither. "It is the mind that moves." The Tibetans also believe that ideas can be realized in physical form, though they qualify this by adding that it would be impossible to conceive any image if the imagined facts did not in the first place correspond to an already-existing external reality.

Alexandra David-Neel, who went to Tibet at the turn of the century and spent fourteen years gathering what is still the most perceptive account of mystic practice there, tells of the technique for generating a phantom or *tulpa* by a powerful concentration of thought. "My habitual incredulity," she says, "led me to make experiments for myself, and my efforts were attended with some success . . . I chose for my experiment a most insignificant character: a monk, short and fat, of an innocent and jolly type. I shut myself in *tsams* and proceeded to perform the prescribed concentration of thought and other rites. After a few months the phantom monk was formed. His form grew gradually fixed and life-like looking. He became a kind of guest, living in my apartment."[123]

Up to this point, there is nothing particularly strange in her story. Many children, even in the West, have imaginary playmates and can tell you at any time exactly where they are and what they happen to be doing. But then Alexandra David-Neel went traveling, and the monk followed her. "Now and then it was not necessary for me to think of him to make him appear. The phantom performed various actions of the kind that are natural to travelers and that I had not commanded . . . he walked, stopped, looked around him." Then, to her dismay, the *tulpa* seemed to escape her control. "The features which I had imagined, when building my phantom, gradually underwent a change. The fat, chubby-cheeked fellow grew leaner, his face assumed a vaguely mocking, sly, malignant look. He became more troublesome and bold." And other people began to see him too, and question her about her friend, the lama. She decided at this stage to dissolve the phantom. "I succeeded," she said later, "but only after six months of hard struggle. My mind creature was tenacious of life."[123]

In the summer of 1917, Elsie Wright and Frances Griffiths, then sixteen and ten years old, spent many days making friends and playing with a community they described as "elves and fairies" in a glen behind their home at Cottingley in Yorkshire. Eventually, to forestall parental disbelief, Elsie borrowed her father's camera and succeeded in taking some fine fairy portraits which were published in *Strand* magazine in 1920.[376] Photographic experts could find no fault with them, and the circumstances in which they were taken seemed to exclude the possibility of fraud. But, since fairies don't exist, reasoned the experts, they must be frauds. So the great creator of master detectives, Sir Arthur Conan Doyle himself, set out with another camera and marked photographic plates. The girls took this equipment to the glen and once again returned with fairy photographs, showing classical little people complete with boots and pointed caps and the traditional gossamer wings. Conan Doyle went on to write a book in which he concluded, "It is hard for the mind to grasp what the ultimate results may be if we have actually proved the existence on the surface of this planet of a population which may be as numerous as the human race, which pursues its own strange life in its own strange way, and which is only separated from ourselves by some difference of vibration."[136]

The material evidence on the photographic plates made it necessary for Conan Doyle, still much in the thrall of Newtonian physics, to emphasize this separation between the phenomena and the consciousness of those involved. Now, in the wake of quantum physical discovery, we have reason to be less certain, less dogmatic about the impossibility of mind affecting matter in this way—even to the extent of being able to set up a thought-form, a *tulpa*, with sufficient reality not only to be seen by other people, but to impose itself on photographic emulsion.

Over a century ago, not long after photography became a

commercial reality, an engraver in Boston discovered that portraits taken of people in his studio included also faint images of others, who turned out to be relatives and friends of the sitters. The fact that some of these ghostly images portrayed people who were already dead led to a belief in "spirit photography" and a continuing field for experiment by camera-carrying mediums.[424] But Ted Serios and others have shown that an image even of something as totally unspiritual as the Denver Hilton Hotel can be impressed on film by someone thinking about it in an appropriate, though as yet unspecified way.[157] So, even if the spiritualists are right about our survival of death and subsequent materialization in photogenic form, it is not necessary to invoke spirits in order to create patterns in the notoriously unstable silver salts on unexposed film. The mind, it seems, can manage on its own. Eusapia Palladino, the extraordinary Neapolitan woman who kept scientists eighty years ago buzzing with speculation, much as Uri Geller does today, was apparently even able on several occasions to produce an imprint of her face or hand in something as substantial as putty kept in sealed containers which she was never allowed to handle in any way.[175]

No perceptual psychologist can tell you exactly what "seeing" means, and few will question that the collative areas in the brain can, on their own, produce patterns which provide an apparently objective sensation—which allows someone to see something that "isn't actually there." And if this phantom vision has sufficient substance to produce mechanical and chemical effects, there is no *a priori* reason why other witnesses present at the same time should not enjoy similar sensations.

In the church of Castelnau-de-Guers in southern France, at Easter in 1974, Abbé Caucanas was kneeling at the altar, his mind filled with thoughts of the Passion, when he saw a face take form on the white napkin covering the Eucharistic bread. When he cried out aloud, his congregation surged forward, and many testified later to seeing a face much like that the abbé described as having "the right eye closed, the left open. The nose was bruised and swollen and bore an expression of pain."[376]

A few reports included blood and tears, and perhaps even a crown of thorns, providing just enough disparity for most critics to dismiss the occurrence as one due to the effects of suggestion. It may have been. I have tried in vain for years, all over India, to see a demonstration of the famous rope trick, and still suspect that its successful performance owes much to hypnotic influence, though I know how hard it is to hypnotize a random group of people which may or may not include a scattering of deep hypnotic subjects. I am aware also of the ease with which an idea can become collective in a crowd, even without any supernatural intrigue.

Under normal circumstances, the left and right halves of the human cerebrum operate in concert. Signals are exchanged very rapidly between them through the connecting corpus callosum, so that perception is apparently simultaneous.[9] An idea arising in the right brain is manifest immediately in the left as well, which for its part has no reason to assume that the notion was conceived in any other place. Transfer of information between the two sides, like the rapid replacement of one frame in a cine film by another, is too fast for us to perceive consciously. Although we each have two relatively independent brains, we are aware of only one identity. So, if and when telepathic transmission occurs between two separate individuals, the chances are that each organism will enjoy the process at firsthand as a subjective personal experience, and not as any kind of intrusion.

Having been involved in a number of experiments to test the possibility of mental communication of information, I have no doubt that something of the sort exists; and I see no reason why it should be confined to a simple two-way telephonic contact, instead of being broadcast in a number of directions. And if it is, or can be, a communal experience, there is little to prevent a number of people gathered together in one place, or widely separated in space, from enjoying the same experience, each in his or her own particular way. In fact, I suspect that much of what we so readily accept as normal everyday experience could well turn out to be rather extraordinary. We take so much for granted, and, perhaps worse, deny so much until it becomes self-evident. The history of science is full of examples of effective, workable systems based on grounds which were only later found to be fallacious. The first kidney transplant was performed in 1952 by a Chicago surgeon on a patient who still had one good functional kidney left. After several months of meticulous observation, he cautiously publicized his report of an apparent success, and because of this almost immediately a rash of similar operations took place all over the world. Later, the Chicago doctor discovered to his horror that the transplant had failed, probably right from the beginning, with the remaining kidney taking on a double load. He published an immediate retraction and apology, but by then nobody cared. Success was universal and has been growing ever since.[427]

I was once taken by a telephone engineer into an automatic exchange where he showed me the mechanics of the switching system and pointed out its limitations. Then almost as an aside he added that in fact, though no one could understand how, the equipment consistently exceeded its limits and handled many more than the theoretical maximum possible number of simultaneous calls. He and his fellow engineers were delighted to take advantage of this unexpected bonus in capacity, and designed all future systems to allow for the inexplicable factor. But he confessed to being rather disturbed by the speed with which he and his associates were knitting together the world's telephone systems, and their associated computers, into one vast, totally automated complex with almost as many units in it as there are cells in a human brain. "I have this recurrent nightmare," he said, "that any day now all the telephones in the world are going to ring simultaneously, and when we answer, there'll be something there that we don't want to hear at all."

This is a purely mechanical example, but at almost every level of physiology we continually accomplish miracles of coordination and perception, about which even the experts among us know little or nothing. There are biological ghosts lurking in every part of life's marvelous machinery. And I sense that one of the most potent of these is the capacity which makes it possible for an idea, strongly held and usually housed at an unconscious level, to manifest its own independent sort of physical reality. How else does one account for the fact that one of the largest blocks of ice ever to fall from the sky came crashing down on April 2, 1973, in Burton Road, Manchester, right at the feet of a physicist concerned with the possibility of ice production by lightning strike?[223] Or for the fact that four species of fish, including a nine-inch largemouth bass *Micropterus salmoides*, came pouring down at breakfast time on October 27, 1947, in Marksville, Louisiana, all around a visiting ichthyologist?[27]

Thomas Bearden, a research physicist with a corporation in Alabama, draws on the "many-worlds" interpretation of quantum mechanics to construct what he calls a "psychotronic" model of reality.[39] This involves complex nuclear mathematics and a tricky extension of Aristotelian logic, but it seems to create a theoretic framework which can accommodate materialization, psychokinetic phenomena, and even antigravity.[38] He shows how it is possible for the personal unconscious to exert physical effects in poltergeist phenomena, and suggests that "the collective species unconscious is vastly more powerful than a personal unconscious, and under appropriate conditions it can directly materialize a thought form, which may be of an object or even of a living being."[40]

Bearden's model allows materialization of any thought, provided sufficient coherence is available to breach the quantum threshold. This could depend on repetition from a single source, or a simultaneous presentation of similar material from a number of minds; either way, the tuning provides appropriate results, giving, among other things, every culture the gods or demons it deserves. As an example, he chooses the Loch Ness monster. "The more intense interest there is, the more photos that are taken, and the more investigation that is done, the easier it is to find evidence of Nessie, because the additional infiltration of the material and the thought form into more and more unconsciousnesses provides more and more tuning stages . . . It appears that Nessie is on the way to being permanently tuned in; a family of plesiosaurs is going to wind up living in Loch Ness, whether or not there are enough fish in the loch to support them."[40]

The same reasoning applies equally well to other paranormal experience. "Since UFO phenomena are unconscious mental forms which are materializing to reality, one can directly 'psychoanalyse' the phenomena." And Bearden makes a very brave attempt to do just that in the following way:

Stalin, he says, made plans during the Second World War to attack the Western countries as soon as they had disarmed sufficiently following the end of conflict with Germany to allow him to do this in relative safety. But his plans were frustrated by the invention of the atomic bomb, and the existence of a small fleet of bombers which imposed balance of power on totally new terms. So, in 1946 and 1947, the grounds for a different type of war, a cold war, were laid in a long-term, worldwide strategy—and the collective unconscious became embroiled in a conflict which subjected large numbers of people in widely separated areas to more or less constant stress. The unconscious tends to draw its symbols from archetypal, prehistoric sources, and, suggests Bearden, in this threatening time the image most in its mind would have been that of the male hunter venturing out from the company of women and children and the shelter of the cave, perhaps with fire in hand, to counter the threat and hold the beasts at bay. Therefore the Soviet threat, that of a dominant male, ought to give rise as it surfaces through the successive layers of the unconscious, where it is modulated and shaped by more personal concerns, to manifest images based on the symbols of fire and the phallus. And, right on cue, fiery ghost rockets begin to be seen in Scandinavian countries in 1947, particularly in those closest to the borders of the Soviet Union.

"But," says Bearden, "the United States is a natural fortress with two great ocean barriers east and west and no strong enemy north or south . . . it corresponds to the cave . . . and thus a female symbol should result there. Bingo! Kenneth Arnold, flying over Washington state—the closest state at the time of the Soviet Union—in 1947 sees flying saucers, which are simply female mandalas modulated by our (science fiction biased) national unconsciousness. Arnold's personal unconscious, the last modulation and shaper of the reality format that emerged, is that of a pilot. So flying saucers are what appeared over the United States."[40]

UFOs in various forms, modulated in different ways by local concerns, became a worldwide phenomenon in the next fifteen years. Then came the 1973 Yom Kippur War. "That," says Bearden, "was the Soviet's 'Spanish Civil War,' where the final tactics and equipment for antitank and antiaircraft weaponry were tested. The effect on the United States and NATO of the cut-off of Arab oil was tested. Further, when the Israelis at one point succeeded in cutting off an Egyptian army, Brezhnev telephoned the United States President and informed him that the Soviets were going in; the United States could do whatever it pleased. The Soviets had seven airborne divisions ready to send in, and the United States had its single airborne division on full alert, ready for insertion. So the world teetered on the brink of direct confrontation between the Soviet Union and the United States, with all its possible consequences." And "confrontation," suggests Bearden, is after all another word for "contact," so it is not surprising that October 1973, the very month the Middle Eastern war erupted, should also be marked by the greatest wave of UFO contactee cases ever to break out in the United States. And, given the recent moonwalks, it was also to be expected that most of the "alien beings" who emerged from their craft on lonely country roads would be wearing spacesuits.[40]

Bearden believes that the next stage in the Soviet program is one based on infiltration, subversion, and a direct attack on the United States using new "psychotronic" weapons. He suggests that this might already have begun, and predicts that if it has, and the collective unconscious is aware of it, there will be a sudden increase in manifest symbols based on an invasion of the cave, on the violation of the female symbol. "The cow is the Western female symbol par excellence . . . so cattle mutilations of a mysterious, paranormal nature have been occurring all across the United States." And just prior to the invasion of NATO itself, says Bearden, "the symbology should be raised to the highest possible degree . . . that is the sexual mutilation of the human female."[40] Perhaps, he suggests, we can already see this happening in the wave of mutilation murders, all concentrating on young women, that took place in Leeds, Boston, and Los Angeles in 1977.

This is an ugly and depressing scenario. Bearden himself says, "The pen, the heart, and the hand that write these words are deeply shaken. For the first time in thirteen years of unrelenting effort, I fervently pray that my insights and analyses might be proved totally in error."[40] He might, on the other hand, be right. His constructs are horribly plausible in their own way, but they do carry a suggestion of the sometimes strained, almost self-fulfilling explanations that come out of superficial Freudian analysis. It is worth bearing in mind that Thomas Bearden, in addition to being a devotee of the new physics, was also once communications officer to a battalion in Korea; director of operations in an area of Vietnam; and chief of a missile intelligence agency in the United States Department of Defense. His long military record, all of it in action against communism, naturally colors his interpretation of the data, drawing only on symptoms of the confrontation between superpowers. These are indeed of great importance in all our affairs, but they can't be all we have on our global minds.

I sense, though, that there is a rightness in his approach. Just as dream content is the natural product of unconscious psychic activity, so paranormal phenomena which seem to be rooted in the collective unconscious should tell us much about its concerns. Jung, as early as 1918, noticed "peculiar disturbances in the unconscious of my German patients which could not be ascribed to their personal psychology." He kept finding there motifs which expressed violence and cruelty. "When I had seen enough of such cases, I turned my attention to the peculiar state of mind then prevailing in Germany. I could only see signs of depression and a great restlessness, but this did not allay my suspicions." He predicted then, twenty years before the actual outbreak of the next world war, that the "blonde beast was stirring in an uneasy slumber" and that an outburst was likely.[297]

Metapsychology, the analysis of group, racial, or national concerns, is still in its infancy. Particular vested interests in business and government, when it seems to their direct commercial or political advantage to do so, consult psychological and psychiatric advisers, much as more antique concerns once sought out the services of a soothsayer.[419] But there is nothing like a United Nations of the Mind, designed to keep a finger on the pulse of the biosphere, ready to swing into global action when alerted by telltale signs in the collective psyche. There could be. We probably already have all the psychological expertise necessary to recognize the symptoms when we see them. All that is missing is the appropriate early warning system, an international network of analysts deployed to monitor sensitive places, paid to take what amounts to perodic polls of opinions prevailing at unconscious levels. And of course a willingness to take these measurements seriously, and to act on the information provided.

The major difficulty facing assessment of this kind is that our present state of consciousness possesses no conceptual categories for unusual experience. It is trained to think of images like UFOs either as physical facts or as metaphysical constructs. Either they are real, or they are not. Either they come from extraterrestrial sources, or they don't exist. We need to change this polar attitude, to recognize that there is a middle ground occupied by psychic factors; by forms inherent in both psychic and physical worlds, which together constitute biological totality.

If a young raven *Corvus corax* is presented with a totally unfamiliar object, it approaches it with all the behavior patterns at its disposal. The bird treats the mystery first as a mortal enemy, launching against it a full-scale attack; then, when this elicits no countermeasures, it regards it as a prize to be killed in the appropriate manner; then the bird switches to dealing with it as a food object, pecking at it, picking it up, tossing it in the air, flying up and dropping it; until finally, when all these techniques prove unavailing, it dismisses the mystery as something indifferent, perhaps merely as an object with which real food may be covered up and eaten.[353] The importance of this broad-fronted approach is that it is strictly experimental. The raven is not necessarily afraid when it flees from the novelty, aggressive when it attacks it, or hungry when it tries to ingest it. It is simply curious.

We humans seem now, however, to have gone some way toward depriving ourselves of the luxury of open curiosity and free experimentation. We have become so adept at ordering sensation that we sometimes organize it out of existence. But that doesn't mean it will simply slip silently away. Psychic entities which are denied access to consciousness in the normal way tend to find some other form of expression, usually in the form of spontaneous projections. These can be, as we have seen, either psychokinetic effects such as those that occur in poltergeistry, diversions in the form of telepathic linkages, or manifestations of apparitions, fairies, monsters, ghosts, or UFOs. All seem to be independent of the individual mind and its nature, but that is almost certainly not true.

Everything in the new physics points to the conclusion that mind and matter share the same dimension in time, and that mind is objective, not subjective. A mental perception of a Loch Ness monster is as real as the monster itself, but lies in a slightly altered "world frame." While it exists only in one observer's mind, in one frame, it is not apparent to other mind frames. But it seems to be possible, though at the moment this is a purely mathematical construct, for two forms to undergo rotation in the same direction so that they superimpose and enjoy the same experience; they "see" the same monster. And any number of others can be included in such a synthesis, until when a sufficiently great number are involved, a consensus is reached through something like the Hundredth Monkey Phenomenon, and the monster becomes a "physical reality." In other words, what we regard as ordinary physical matter is simply an idea that occupies a world frame common to all minds. The universe is literally a collective thought, and we have a very powerful say in the reality manifest in our particular sector.

I suggest that what we have in the field of paranormal phenomena are vivid and sometimes meaningful indicators of the true state of our psyche. The different physical realities of raps on tables, bent keys, flying saucers, and wandering yetis are in themselves purely incidental symbols, which may sometimes coincide in disturbing ways, but which bear no recognizable causal relationship to one another. We must not expect scattered occurrences, stray scraps of information, to carry any information of consequence. Each on its own says very little, and any attempt to wring an answer from isolated experience, or to massage meaning out of laboratory statistics, is doomed to failure, as is any analytic endeavor which tries to operate under the restrictive umbrella of any particular cult or persuasion, no matter whether it be total scientific objectivity, a belief in extraterrestrial intervention, or an insistence on the mediation of friendly spirits.

The stuff of parapsychology, supernature, and the occult is partly flotsam, thrown up to the surface of life by currents and eddies in the Lifetide; and partly jetsam, abandoned to the tide by ever more complex organisms intent on lightening their collective loads. Some of it is useful and significant, and some is bound to be senseless and absurd. And in all our dealings with any of this salvage, it would be well to bear in mind Jung's observation: "The best, just because it is the best, holds the seed of evil, and there is nothing so bad but good can come of it."[298] The world, by and large, is what we make of it.

As a biologist, aware of the breadth and extent of life, and conscious of many of its persistent mysteries, I am grateful for any scraps of understanding. I believe that recent advances in astrophysics at one end of the scale, and molecular biology and genetics at the other, put us now in a position to begin to make reasoned judgments about the general process of evolution which has led to our present, strange, subdivided state.

It seems to me that the seeds of life were formed in, and broadcast from, interstellar space. Many took root here and grew, with the aid of patterns in our soil, into a unique earthbound biosphere. Control of this biosphere is largely in the hands of the first replicators, the genes, which find selective advantage in forming communities. Many of these associations, by virtue of their special form, have acquired an individuality which is partly independent of the genes. In some, this individuality has grown into an awareness of communal identity which enjoys sufficient freedom from the genes to produce marked internal conflict. And it is such conflict which, in a very few organisms, has resulted in the ultimate liberation of consciousness.

But that is by no means the end of the story. Total emancipation from instinctive genetic domination is still curtailed by the existence of tidal influences which in turn have a long evolutionary history. I have dubbed these forces the contingent factors and believe that they have played a major part in freeing us, but that they are influences over which we still have little or no control. I am by no means persuaded that we ought even to seek complete liberation from these contingencies. Inherent in them there seems to be a natural order, almost a basic cosmic rhythm, which probably underlies all coincidence, chance, seriality, and synchronicity.[414] I believe we will eventually succeed in laying this ghost in our machine, and giving it all the physical properties and parameters necessary for its establishment as a recognized force of nature. And when this is accomplished, I feel certain that it will be seen to be the source of much that we now regard as supernatural. But I don't expect even such revelations to provide the answer to the basic mysteries of life and mind.

These, I am sure, are eternal intangibles. They have their

roots in organic evolution, but they have no more substance than the tide. You can collect as many seawater samples as you like, but none will contain, nor tell you anything about, the tide. You can dissect as many living organisms as you can lay your hands on, breaking them down into their subatomic components, and still find no answers. Life is a pattern, a movement, a syncopation of matter; something produced in counterpoint to the rhythms of contingency; a rare and wonderfully unreasonable thing.

# THE TIDEMARK: A Review

I am conscious, here near the end, of having failed at times to find the right words with which to express something that may anyway prove to be ineffable. Words, with their weight, have a tendency to fall like birds of prey on delicate ideas, carrying them away before they have a chance to reach fruition. I have been forced, for instance, at several critical junctures, to fall back on the metaphor of the tide.

Julian Jaynes would see in this no reason for apology. "Understanding a thing is to arrive at a metaphor for that thing by substituting something more familiar to us. And the feeling of familiarity is the feeling of understanding." He believes that imaginative use of metaphor is the very ground of language, and that it was this use of language which made us conscious beings. As a biologist, I must invert that causal chain and suggest that we would never have been able to make much of a metaphor had we not already been manifestly conscious. However, we arrive in the end at a similar conclusion, which is that "Mind is an analog of what is called the real world."[284]

The Soviet psychologist Aleksandr Luria suggests that words, far from being a simple shorthand for communication, actually structure and reorganize reality. "As a race of people share the same language, they share the same reality."[356] A word which designates an object carves out and fixes it, giving it a defined status quite independent of its function. In all European languages, there is a predominance of nouns, whose function is largely static. "Whatever is designated by a noun becomes a thing, a substance with properties; and for speakers whose vocabularies consist as largely of nouns as ours do, the world consists mainly of physical objects."[334] We can see a Cheshire cat with or without a grin, but not a grin without a cat. For, in our language, "a grin" is an act, which requires an agent, which must be a permanent entity. We literally talk ourselves out of a whole range of possible experience.

In Australian languages there are no "parts of speech" comparable to ours, but the main words are ones of action, rather like our verbs. These do not order things into rigid categories, or insist on objects as the fixed focus of experience. They describe instead a whole series of related acts, with conjugations of each word expressing its state of completion, and not its time of performance. Construction of this kind frees an agent from the strict and purely metaphysical logic of our language, which insists that things must be either true or false, real or imaginary. We can only escape these rigid bounds in mathematics or deliberate "nonsense," but for an aborigine raised in the Arunta tongue, a grin may be grinned by a Cheshire cat, or a totem cat, or by no cat at all; because he has no fixed and predetermined concept of a "cat."[499]

I would like my concept of the Lifetide to be a little like this, most clearly conveyed perhaps by thinking of "tide," not as a noun, but as an intransitive verb ungoverned by any object. Though even with this enlarged perspective, I am still very much aware of the sheer tyranny of words.

In early 1974, "a slow-moving, solidly built little girl with a distant smile" was brought by her mother to the Child Development Research Unit at Nottingham University. Nadia was six and still virtually mute, large for her age, lethargic and poorly coordinated. She had been unresponsive to her mother as a baby, and continued to show so little reaction to the usual social approaches that she was soon diagnosed as autistic and sent to a school for "severely subnormal children." In playrooms filled with toys and delights enough to amaze most children, she remained passive and unmoved, unless she was involved in one of her rare and uncontrollable bouts of screaming.[551]

On the first day at the unit, psychologists sat with Nadia's mother, watching the child through a one-way screen, discussing her history and preparing to deal with her as they do each day with the dozens of retarded children in their care. Nadia "had been presented with the fat wax crayons which seemed suitable to the general level of play and manipulative competence that she had shown so far, and was scrubbing away at the paper to produce a formless yellow scribble . . . We started to talk about the things Nadia liked doing at home, and her mother rummaged in her bag and produced a bundle of half-a-dozen drawings. I looked at them in the half-light: rather faint blue lines on poor quality paper. I went through the little pile again . . . My first reaction to the drawings was to marvel; my second, I am ashamed to say, to doubt . . . the drawings were not possible."[406]

Incredulity is the usual response to Nadia's drawings. From the age of three, when she was first given access to her preferred medium, a ballpoint pen, she was able to produce line drawings of extraordinary sophistication. She would not draw to order, but only when moved to do so. Then, left-handed, very swiftly and deftly, she composed with fine, certain strokes. "Her lines were firm and executed without unintentional wavering. She could stop a line exactly where it met another despite the speed with which the line was drawn. She could change the direction of a line and draw lines at any angle towards and away from the body. She could draw a small and perfect circle in one movement and place a small dot in the centre."[483] Given a ballpoint pen, this "clumsy and retarded" child was transformed into a skilled representational artist, producing drawings rich with life and movement, showing a complete mastery of perspective and foreshortening. Her horses and riders leap out of the page at you at angles which would tax the skill of any adult artist. Her crowing cockerels are vibrant with early-morning vitality. Her footballers are poised, caught in mid-action, with astonishing sensitivity. Nadia breaks every rule there is about how we perceive, conceive, and learn to draw.

It has been said, "Art is born of art—not of nature," and that all paintings owe more to other paintings than they do to observation.[208] But Nadia draws directly from life, or gets her inspiration from crude wash drawings in cheap picture books, and seems to process the images in her head so that what comes out is not a photographic representation of a scene, or a copy of an illustration, but a living structure which grows and changes each time she portrays it. She interprets nature in an almost aboriginal way, as though she too were dealing in active, verbal images instead of passive nouns and diagrams. And she is capable of sharing this vision with others. Or she was.

In 1975, Nadia's mother died and she was admitted to a special school for autistic children, where intensive, well-meaning efforts are being made to teach her how to speak; how to conform to our description of reality. Slowly, step by step, she is beginning to use the right and expected words, and, as her language improves, so her artistic abilities are smothered. The headmaster of the school says with satisfaction, "We get far more speech and socialisation out of her now she has stopped drawing as a matter of course."[85] From a child with an unique expressive talent that gave her, in tests which assess intelligence from drawing ability, an enormous IQ of 160 when she was only four, Nadia has been molded into a tragic, subnormal ten-year-old, who will probably never be able to meet life successfully on a verbal basis, or earn her own livelihood in any way. All that is left of an extraordinary talent which might, given the right response from our society, have taught us a great deal about Nadia and about ourselves is an occasional ghostly sketch of a horse and rider that appears on the steamed-up windows of her school, when no one in authority is watching.

Our entire educational effort seems to be dedicated to making others over in our image, turning out carbon-copy people who faithfully reproduce all our mistakes. "Better," we seem to be saying, "a poor imitation of ourselves, than something different, something threatening." An eminent child psychologist sums up Nadia's case by concluding that if the loss of her gift is "the price that must be paid for language—even just enough language to bring her into some kind of community of discourse with her small protected world—we must, I think, be prepared to pay that price on Nadia's behalf."[406]

Must we? I am far from convinced. I value language very highly, but not above all else. I am greatly concerned about the way we treat the special ones, the strange ones who seem to be able to respond in some way to the underlying form of things, instead of getting stuck as we do on surface details. They are closer to the roots of being, more involved than the rest of us, more touched by continuity. They respond to direct sensation, they dance not to our monotonous tune, but to the light, drawing inspiration from the earth itself. Instinct tells me that if we are to make any real sense of the universe, we need the ones who see and do things differently—the child prodigies who begin composing music almost before they can walk; the "idiot savants" who cannot read or write, but can work out square roots to the fifth decimal place in seconds; the metal benders and the visionaries. They rise, like repressed archetypes, directly from the unconscious, reminding us of its potential. I believe they could all teach us something, if we let them. But we so seldom do.

It really is a matter of maintaining balance. We have temporarily lost ours by coming down too heavily in one direction, on the side of technology and reason, and we have begun to pay the price. It's probably not too late to change, but we can't waste any more time or pass up any further opportunities. They have been coming thick and fast in the last five years, but if my own experience is anything to go by, the tide may be just about to turn.

I went to Venice again this year, to visit the family who live near Santa Maria dei Miracoli. Claudia is eight now and goes by herself to school. In the breaks between their classes, she and the other children play around the old well head in the *piazzetta*, practicing the lessons they have learned, confirming the consensus. Sometimes they throw and catch a tennis ball, letting it bounce and roll between them; and now that it has a name and its function is fixed and defined, it doesn't even occur to Claudia to offer it any other kind of freedom. She is one of us.

It would be sad, and could be fatal, if we continue to allow all the clarity of childish perception to be clouded in this way. For there seem to be moments, before we become familiar with the house rules, when we are able to see right through the cracks in the cosmic egg and almost touch the truth.

So I dedicate this book to all those who still feel the living tug of the tide, and who have the courage to let it carry them wherever it may go.

# Bibliography

1. AHMADJIAN, V. " The fungi of lichens," *Scientific American* *208*: 122–132, 1963.
2. AHMADJIAN, V. "Artificial reestablishment of the lichen *Cladonia cristatella*," *Science 151*:199–201, 1966.
3. AHMADJIAN, V. "Lichens," in HENRY (249).
4. ALLEN, L. A. *Time before morning*. New York: Thomas Crowell, 1975.
5. ALTMANN, S. A. (ed.) *Social communication among primates*. University of Chicago Press: Chicago, 1967.
6. ALVAREZ, F., et al. "Experimental brood parasitism of the magpie," *Animal Behaviour 24*: 907–916, 1976.
7. ANDERS, E., & FITCH, F. W. "Search for organized elements in carbonaceous chondrites," *Science 138*: 1392–1399, 1962.
8. ANDERS, E., et al. "Organic compounds in meteorites," *Science 182*: 781–790, 1973.
9. ANDREW, K. "Psychokinetic influences on an electromechanical number generator . . ." in MORRIS et al. (386).
10. ANNIS, R. C., & FROST, B. "Human visual ecology and orientation anisotropies in acuity," *Science 182*: 729–731.
11. ANON. National Geographic Society Press Release, March 31, 1955.
12. ANON. Detroit *Free Press*, December 14, 1959.
13. ANON. *Malay Mail*, Kuala Lumpur, August 15, 1966.
14. ANON. *Extraterrestrial life: An anthology and bibliography*. United States National Academy of Sciences: Washington, 1966.
15. ANON. "The effects of UFOs upon animals," *Flying Saucer Review 16*: 1, 1970.
16. ANON. Graffito at the University of Michigan, 1970.
17. ANON. *The Times*, London, September 24, 1973.
18. ANON. Personal files, October 4, 1974.
19. ANON. *Scientific American 237*: 70–72, 1977.
20. ANON. *Newsweek*, New York, January 15, 1978.
21. ANTROBUS, J. S., et al. "Experiments accompanying daydreaming," *Journal of Abnormal and Social Psychology 69*: 244–252, 1964.
22. ANTROBUS, J. S., et al. "Studies in the stream of consciousness," *Perceptual and Motor Skills 23*: 399–417, 1966.
23. ASERINSKY, E., & KLEITMAN, N. "Regularly occurring periods of eye motility and concomitant phenomena during sleep," *Science 118*: 273–274, 1953.

24. AYALA, F. J., & DOBZHANSKY, T. '*Studies in the philosophy of biology.*' Berkeley: University of California Press, 1974.
25. BADDELEY, A. D. *The psychology of memory.* New York: Harper & Row, 1976.
26. BAERENDS, G., et al. (eds.) *Essays on function and evolution in behaviour.* Oxford: Clarendon Press, 1975.
27. BAJKOV, A. D. "Do fish fall from the sky?" *Science 109*: 402, 1949.
28. BANKAN, P. "Hypnotizability, laterality of eye movements and functional brain assymetry," *Perceptual and Motor Skills 28*: 927–932, 1969.
29. BALL, G. H. "Organisms living on and in protozoa," in CHEN (105).
30. BALTENSWELLER, W. "*Zeirapheira griseana* in the European Alps . . ." *Canadian Entomologist 96*: 792–800.
31. BARBER, T. X. "Experimental evidence for a theory of hypnotic behavior," *International Journal of Clinical and Experimental Hypnosis 9*: 181–193, 1961.
32. BARBER, T. X. "Hypnotic age regression: A critical review," *Psychosomatic Medicine 24*: 286–299, 1962.
33. BARBER, T. X. *Hypnosis: A scientific approach.* New York: Van Nostrand, 1969.
34. BARGHOORN, E. S., & SCHOPF, J. W. "Microorganisms three billion years old from the Precambrian of South Africa," *Science 152*: 758–763, 1966.
35. BATCHELDOR, K. J. "Report on a case of table levitation and associated phenomena," *Journal of the Society for Psychical Research 43*: 339–356, 1966.
36. BATCHELDOR, K. J. *Practical hints for small group study of PK using tables.* Privately published: c/o 1 Adam & Eve Mews, London.
37. BATES, D. R. (ed.) *The planet Earth.* Oxford: Pergamon, 1964.
38. BEARDEN, T. E. *A postulated mechanism that leads to materialization* . . . United States Army Missile Command, Technical Report SAM-D 76–1, 1975.
39. BEARDEN, T. E. *An approach to understanding psychotronics.* Privately published: 1902 Willis Road, Huntsville, Ala., 1977.
40. BEARDEN, T. E. *Species metapsychology, UFO waves, and cattle mutilations.* Privately published, 1977.
41. BEARDEN, T. E., Personal communication, December 8, 1977.
42. BEATTIE, J. "Divination in Bunyoro, Uganda," in J. Middleton, (ed.), *Magic, witchcraft, and curing.* New York: Natural History Press, 1967.
43. BEAUMONT, P. B. "The ancient pigment mines of Southern Africa," *South African Journal of Science 69*: 140–146, 1973.
44. BEAUMONT, P. B. "Border Cave—a progress report," *South African Journal of Science 69*: 41–46, 1973.
45. BEER, C. G. "Multiple functions and gull displays," in BAERENDS (26).

46. BENNETT, E. *Apparitions and haunted houses.* London: Faber, 1939.
47. BENVENISTE, R. E., & TODARO, G. J. "Evolution of C-type viral genes . . . *Nature* 252: 456–458, 1974.
48. BERGSON, H. L. *Matter and memory.* London: Allen & Unwin, 1911.
49. BERGSON, H. L. Presidential Address. *Proceedings of the Society for Psychical Research* 26: 462–479, 1913.
50. BERGSON, H. L. *The two sources of mortality and religion.* New York: Holt, 1935.
51. BERNAL, J. D. *The origin of life.* London: Weidenfeld & Nicolson, 1965.
52. BERNE, E. *A layman's guide to psychiatry and psychoanalysis.* London: Andre Deutsch, 1969.
53. BERNSTEIN, M. *The search for Bridey Murphy.* New York: Doubleday, 1956.
54. BERTRAM, B. C. R. "The vocal behaviour of the Indian Hill Mynah," *Animal Behaviour Monograph No. 3*, 1970.
55. BETTLEY, F. R. "Ichthyosis and hypnosis," *British Medical Journal* (2): 996, 1952.
56. BIRDSELL, J. B. *Human evolution.* Chicago: Rand McNally, 1972.
57. BLACK, S. "The use of hypnosis in the treatment of psychosomatic disorders," *Proceedings of the society for Psychosomatic Research Conference of 1962.* Oxford: Pergamon, 1964.
58. BLACK, S. *Mind and body.* London: William Kimber, 1969.
59. BLAIR, L. *Rhythms of vision.* London: Croom Helm, 1975.
60. BLEEK, W. H. I., & LLOYD, L. C. *Specimens of Bushman folklore.* London: George Allen, 1911.
61. BOAG, T. J., & CAMPBELL, D. (eds.). *A triune concept of the brain and behavior.* Toronto: University of Toronto Press, 1973.
62. BOARDMAN, R. S., et al. (eds.). *Animal societies.* Dowden, Hutchinson & Ross, 1973.
63. BOHM, D. "Some remarks on the notion of order," in WADDINGTON (565).
64. BONNER, J. T. "How slime molds communicate," *Scientific American* 209: 84–93, 1963.
65. BONNER, J. T. *Cells and societies.* Princeton, N.J.: Princeton University Press, 1965.
66. BONNER, J. T. *The cellular slime molds.* Princeton, N.J.: Princeton University Press, 1967.
67. BONNER, J. T. "The chemical ecology of cells in the soil," in SONDHEIMER & SIMEONE (496).
68. BOSHIER, A. K. "Ancient mining of Bomvu Ridge," *Scientific South Africa* 2: 317–320, 1965.
69. BOSHIER, A. K. *Mining genesis.* Johannesburg: Mining Survey, 1970.

70. Boshier, A. K. Personal communication, 1977.
71. Boshier, A. K., & Beaumont, P. B. *Mining in southern Africa and the emergence of modern man.* Johannesburg: Optima, 1972.
72. Bowen, C. *The humanoids.* London: Neville Spearman, 1969.
73. Bowers, K. S. *Hypnosis for the seriously curious.* Monterey, Calif.: Brooks-Cole, 1976.
74. Bowlby, J. *Attachment and loss.* (2 vols.). London: Hogarth, 1969.
75. Breznak, J. A. "Symbiotic relationship between termites and their intestinal microbiota," in Jennings & Lee (286).
76. Broadbent, D. E. *Perception and communication.* Oxford: Pergamon, 1958.
77. Brodeck, M. "Mental and physical," in Feyerbend & Maxwell (172).
78. Brody, E. B. "Research in reincarnation and editorial responsibility," *Journal of Nervous and Mental Disease 165*: 151, 1977.
79. Bronowski, J., & Bellugi, U. "Language, name and concept," *Science 168*: 669–673, 1970.
80. Brooks, J., & Shaw, G. "Evidence for extraterrestrial life," *Nature 223*: 754–756, 1969.
81. Brookes-Smith, C., & Hunt, D. W. "Some experiments in psychokinesis," *Journal of the Society for Psychical Research 45*: 265–281, 1970.
82. Brown, B. *New mind, new body.* London: Hodder & Stoughton, 1975.
83. Brown, M. H. *PK.* Blauvelt, N.Y.: Steinerbooks, 1976.
84. Bruner, J. *On knowing.* Cambridge, Mass.: Harvard University Press, 1962.
85. Bugler, J. "The genius of Nadia," *Observer* magazine: London, November 27, 1977.
86. Burma, B. H. 'The species concept . . .' *Evolution 3*: 369–370, 1949.
87. Burnet, F. M. 'The evolution of adaptive immunity in vertebrates,' *Acta. Pathology and Microbiology, Scandinavia 76*: 1–11, 1969.
88. Burnet, F. M. *Immunological surveillance.* Oxford: Pergamon, 1970.
89. Burnet, F. M. "Self-recognition in colonial marine forms and flowering plants," *Nature 232*: 230–235, 1971.
90. Busnel, R.-G. "Acoustic communication," in Sebeok (481).
91. Cairns-Smith, A. G. "The origin of life and the nature of the primitive gene," *Journal of Theoretical Biology 10*: 53–88, 1966.
92. Cairns-Smith, A. G. *The life puzzle.* Edinburgh: Oliver & Boyd, 1971.
93. Cairns-Smith, A. G. "A case for alien ancestry," *Proceedings of the Royal Society of London B 189*: 249–274, 1975.

94. CAIRNS-SMITH, A. G. "Synthetic life for industry," in DUNCAN & WESTON-SMITH (141).
95. CALDER, R. *Man and the cosmos*. London: Pall Mall Press, 1968.
96. CAMERON, A. G. W. "Clumping of interstellar grains . . ." *Icarus 24*: 128–133, 1975.
97. CAMPBELL, B. *Sexual selection and the descent of man*. Aldine: Chicago, 1972.
98. CANETTI, E. *Crowds and power*. London: Victor Gollancz, 1962.
99. CAPRA, F. *The Tao of physics*. London: Wildwood House, 1975.
100. CARPENTER, C. R. "The howlers of Barro Colorado Island," in DE VORE (130).
101. CARPENTER, E. *Oh, what a blow that Phantom gave me!* New York: Holt, Rinehart & Winston, 1973.
102. CATHIE, B. *Harmonic 695*. Wellington: A. H. & A. W. Reed, 1971.
103. CAVALIER-SMITH, T. "The origin of nuclei and of eukaryotic cells," *Nature 256*: 463–468, 1975.
104. CAYCE, E. *Dreams: The language of the unconscious*. Virginia Beach: A.R.E. Press, 1962.
105. CHEN, T. T. *Research in protozoology*. Oxford: Pergamon, 1969.
106. CHERTOK, L. *Pyschophysiological mechanisms of hypnosis*. Berlin: Springer, 1969.
107. CHETWYND, T. *Dictionary for dreamers*. London: Allen & Unwin, 1972.
108. CHOW, L. T., et al. "An amazing sequence arrangement at the 5' ends of adenovirus 2 messenger RNA," *Cell 12*: 1–8, 1977.
109. CHRISTOPHER, M. *Seers, psychics and ESP*. London: Cassell, 1970.
110. CLARK, F., & SYNGE, R. L. M. The origin of life on Earth. Oxford: Pergamon, 1959.
111. CLARK, L. R., et al. *The ecology of insect populations in theory and practice*. London: Methuen, 1967.
112. CLAUS, G., & NAGY, B. 'A microbiological examination of some carbonaceous chondrites,' *Nature 192*: 594–596, 1961.
113. COLE, G. D. H. (ed.). *The essential Samuel Butler*. London: Everyman, 1950.
114. COLERIDGE, S. T. *Poetical works*. Oxford: Oxford University Press, 1912.
115. COOKE, H. B. S., et al. "Fossil man in the Lebombo Mountains, South Africa," *Man 3*: 6–13, 1945.
116. COULSON, J. C. "The influence of change of mate on the breeding biology of the kittiwake," *Animal Behaviour 14*: 189–190, 1966.

117. CRASILNECK, H. B., & HALL, J. A. "Physiological changes associated with hypnosis," *Journal of Clinical and Experimental Hypnosis* 7: 9–50, 1959.
118. CREIGTON, G. "The humanoids in Latin America," in BOWEN (72).
119. CUDMORE, L. L. L. *The center of life.* New York: Quadrangle, 1977.
120. CULLEN, J. "Reduction of ambiguity through ritualization," in HUXLEY (274).
121. CYNADER, M. & MITCHELL, D. "Monocular astigmatism effects on kitten visual cortex development," *Nature 270*: 177–178, 1977.
122. DARWIN, C. *On the origin of species.* (1859) Harmondsworth, England: Penguin, 1968.
123. DAVID-NEEL, A. *Magic and mystery in Tibet.* (1931) London: Souvenir, 1967.
124. DAVIDSON, E. H. *Gene activity in early development.* New York: Academic Press, 1968.
125. DAWKINS, R. *The selfish gene.* Oxford: Oxford University Press, 1976.
126. DE BEER, G. R. *Embryos and ancestors.* Oxford: Oxford University Press, 1951.
127. DEMENT, W. C. Discussion on paper by SNYDER (495).
128. DE ROCHAS, A. *Les vies successives.* Paris, 1911.
129. DEVEREUX, G. *Psychoanalysis and the occult.* London: Souvenir, 1974.
130. DE VORE, I. (ed.). *Primate behavior.* New York: Holt, Rinehart & Winston, 1965.
131. DOBZHANSKY, T. *The biology of ultimate concern.* New York: New American Library, 1967.
132. DOETSCH, R. N., & COOK, T. M. *Introduction to bacteria and their ecobiology,* Baltimore: University Park Press, 1973.
133. DOUGLAS, A. E. "Origins of diffuse interstellar lines," *Nature 269*: 130–132, 1977.
134. DOUGLAS, M. *Tales and legends of Northumbria.* London: Houghton & Scott-Snell, 1934.
135. DOUGLAS, N. *The Book of Matan.* Sudbury, England: Neville Spearman, 1977.
136. DOYLE, A. C. *The coming of the fairies.* London: Hodder and Stoughton, 1922.
137. DRIESCH, H. *The history and theory of Vitalism.* London: Macmillan, 1914.
138. DUCASSE, C. *A critical examination of the belief in a life after death.* Springfield, Ill.: Charles Thomas, 1960.
139. DUNCAN, C. J. *The molecular properties and evolution of excitable cells.* Oxford: Pergamon, 1967.
140. DUNCAN, C. J. "A note on the evolution of the transducer mechanism of the vertebrate retinal rod," *Experientia 33*: 1310, 1977.

141. DUNCAN, R., & WESTON-SMITH, M. *The encyclopaedia of ignorance*. Pergamon: Oxford, 1977.
142. DUNDES, A. "Science in folklore? Folklore in science?" *New Scientist* 76: 774–776, 1977.
143. ECCLES, J. "The experiencing self," in ROSLANSKY (446).
144. ECCLES, J. (ed.). *Brain and conscious experience*. Berlin: Springer, 1966.
145. EDDINGTON, A. S. *New pathways in science*. Cambridge: Cambridge University Press, 1935.
146. EDMONSTON, W. E., & PESSIN, M. "Hypnosis as related to learning and electrodermal measures," *American Journal of Clinical Hypnosis* 9: 31–51, 1966.
147. EDMUNDS, S. *Hypnotism and the supernormal*. London: Aquarian Press, 1961.
148. EDWARDS, F. *Strangest of all*. London: Pan, 1962.
149. EHRENWALD, J. "Mother-child symbiosis," *Psychoanalytic Review* 58: 455–466, 1971.
150. EHRENWALD, J. "The telepathy hypothesis and schizophrenia," *Journal of the American Academy of Psychoanalysis* 2: 159–169, 1974.
151. EHRENWALD, J. "Cerebral localization and the psi syndrome," *Journal of Nervous and Mental Disease* 161: 393–398, 1975.
152. EHRENWALD, J. "Therapeutic applications," in KRIPPNER (327).
153. EIBL-EIBESFELDT, I. *Ethology*. New York: Holt, Rinehart & Winston, 1970.
154. EINSTEIN, A. *The world as I see it*. London: John Lane, 1935.
155. EISELEY, L. *Coming of the giant wasps*. Audubon, 1975.
156. EISENBUD, J. "Psychiatric contributions to parapsychology," in DEVEREUX (129).
157. EISENBUD, J. *The world of Ted Serios*. London: Jonathan Cape, 1968.
158. EISENBUD, J. Appendix to ULLMANN et al. (552).
159. ELIADE, M. *Myths, dreams, and mysteries*. New York: Harper & Row, 1960.
160. ELIADE, M. *Myth and reality*. New York: Harper & Row, 1963.
161. ELIADE, M. *The forge and the crucible*. New York: Harper & Row, 1971.
162. ELLIS, K. *Science and the supernatural*. London: Wayland, 1974.
163. ELIOT, T. S. *Ash Wednesday*. New York: Harcourt, Brace, 1930.
164. ELSASSER, W. M. *Atom and organism*. Princeton, N.J.: Princeton University Press, 1966.
165. EPHRUSSI, B., & WEISS, M. C. "Hybrid somatic cells," *Scientific American* 220: 26–31, 1969.
166. ESTABROOKS, G. H. *Hypnotism*. New York: E. P. Dutton, 1957.
167. EVANS, F. J., et al. "Sleep-induced behavioural response," *Journal of Nervous and Mental Disease* 148: 467–476, 1969.

168. EVANS, H. E. *Life on a little-known planet.* New York: E. P. Dutton, 1966.
169. EVANS-PRITCHARD, E. E. *Witchcraft, oracles and magic among the Azande.* Oxford: Clarendon Press, 1937.
170. FABRE, J. H. *The hunting wasps.* (1879) New York: Dodd, Mead, 1915.
171. FEINBERG, G. "Precognition: A memory of things future," in OTERI (417).
172. FEYERABEND, P. K., & MAXWELL, G. (eds.). *Mind, matter and method.* Minneapolis: University of Minnesota Press, 1966.
173. FISKE, D. & MADDIS, S. (eds.). *Functions of varied experience.* Dorsey Press: Homewood, Ill., 1961.
174. FISS, H., et al. "Waking fantasies following interruptions of two types of sleep," *Archives of General Psychiatry 14:* 543–551, 1966.
175. FLAMMARION, C. *Mysterious psychic forces.* Boston, 1907.
176. FODOR, N. *Freud, Jung and occultism.* New York: University Books, 1971.
177. FORDHAM, F. *An introduction to Jung's psychology.* Harmondsworth, England: Penguin, 1966.
178. FORT, C. *The books of Charles Fort.* New York: Holt, Rinehart & Winston, 1941.
179. FOSTER, D. *The intelligent universe.* London: Abelard, 1975.
180. FOULKES, D. "Theories of dream formation," *Psychological Bulletin 62:* 236–247, 1964.
181. FOULKES, D. *The psychology of sleep.* New York: Scribner's, 1966.
182. FOX, A. S., & VALENCIA, J. I. "Gene transfer in *Drosophila melanogaster,*" *Genetics 53:* 897–911, 1975.
183. FOX, S. W. (ed.). *The origins of prebiological systems.* New York: Academic Press, 1965.
184. FRANKLYN, J. *Death by enchantment.* London: Hamish-Hamilton, 1971.
185. FRANTZ, R. "Pattern vision in newborn infants," *Science 140:* 296, 1963.
186. FREEDMAN, A. M., et al. *Comprehensive textbook of psychiatry.* Baltimore: Williams & Wilkins, 1975.
187. FREEMON, F. R. *Sleep research.* Springfield, Ill.: C. C. Thomas, 1972.
188. FREUD, S. *The interpretation of dreams.* (1900) Standard Edition 4 and 5. London: Hogarth Press.
189. FREUD, S. *Totem and taboo.* (1913) Standard Edition 13.
190. FREUD, S. *Dreams and telepathy.* (1925) Standard Edition 4.
191. FREUD, S. *The future of an illusion.* (1927) Standard Edition 20.
192. FREUD, S. *New introductory lectures on psychoanalysis.* (1933) Standard Edition 22.
193. FREUD, S., & BREUER, J. *Studies on hysteria* (1895) Standard Edition 2.

194. GARDNER, R. A., & GARDNER, B. "Teaching sign language to a chimpanzee," *Science 165*: 664–672, 1969.
195. GARDNER, R. A. & GARDNER, B. "Two way communication with an infant chimpanzee," in SCHRIER & STOLLNIZ (479).
196. GASTAUT, H., & BROUGHTON, R. "A clinical and polygraphic study in episodic phenomena during sleep," in WORTIS (599).
197. GAZZANIGA, M. S. "The split brain in man," *Scientific American 217*: 24–29, 1967.
198. GEIST, V. "The evolutionary significance of mountain sheep horns," *Evolution 20*: 558–566, 1966.
199. GIBSON, H. B. *Hypnosis*. London: Peter Owen, 1977.
200. GIBSON, G. E., & GIAUQUE, W. F. "The third law of thermodynamics . . ." *Journal of the American Chemical Society 45*: 93–97, 1923.
201. GIDRO-FRANK, L., & BOWERSBUCH, M. K. "A study of the plantar response in hypnotic age regression," *Journal of Nervous and Mental Disease 107*: 443–458, 1948.
202. GILL, M. M. "Hypnosis as an altered or regressed state," *International Journal of Clinical and Experimental Hypnosis 20*: 224–237, 1972.
203. GILLHAM, N. W. *Organelle heredity*. New York: Holt, Rinehart & Winston, 1976.
204. GILLIARD, E. T. "The Evolution of Bowerbirds," *Scientific American 209*: 38–46, 1963.
205. GILLIARD, E. T. *Birds of paradise and bowerbirds*. London: Weidenfield & Nicholson, 1969.
206. GLASSER, R. J. *The body is the hero*. London: Collins, 1977.
207. GOLDANSKII, V. I. "Interstellar grains as possible cold seeds of life," *Nature 269*: 583–584, 1977.
208. GOMBRICH, E. H. *Art and illusion*. New York: Pantheon, 1960.
209. GOOCH, S. *The Neanderthal question*. London: Wildwood House, 1977.
210. GOODALL, J. "The behaviour of free-living chimpanzees . . ." *Animal Behaviour Monographs No. 1*: 161–311, 1968.
211. GOODWIN, D. W. "Alcohol and recall," *Science 163*: 1358–1360, 1969.
212. GORDON, H., & COHEN, K. "Case of congenital linear naevus treated by hypnosis," *International Dermatology Congress 10*: 376, 1952.
213. GOTTFRIED, B. S., et al. "The Leidenfrost phenomenon," *International Journal of Heat and Mass Transfer 9:* 1167–1187, 1966.
214. GRANT, J. *Winged pharaoh*. London: Victor Gollancz, 1937.
215. GRAVES, T. *Dowsing*. London: Turnstone, 1976.
216. GREEN, C., & McCREERY, C. *Apparitions*. London: Hamish-Hamilton, 1975.
217. GREENBERG, M., & HULST, H. C. van de (eds.). *Interstellar dust and related topics*. Dordrecht: Reidel, 1973.
218. GREENBERG, J. M. "The interstellar depletion mystery . . ." *Astrophysical Journal 189*: L81–L85, 1974.

219. GREENHOUSE, H. B. *Premonitions.* London: Turnstone, 1971.
220. GREGORY, R. L. *The intelligent eye.* London: Weidenfeld & Nicolson, 1970.
221. GREYSON, B. "Telepathy in mental illness," *Journal of Nervous and Mental Disease 165*: 184–200, 1977.
222. GRIFFIN, D. R. *The question of animal awareness.* New York: Rockefeller University Press, 1976.
223. GRIFFITHS, R. F. "Observation and analysis of an ice hydrometeor of extraordinary size," *Meteorological Magazine 104*: 253–260, 1975.
224. GRIMBLE, A. *A pattern of islands.* London: John Murray, 1952.
225. GRIMBLE, A. *Migrations, myth and magic from the Gilbert Islands.* London: Routledge & Kegan Paul, 1972.
226. GRIMSTONE, A. V., & CLEVELAND, L. R. "The fine structure and function of the contractile axostyles of certain flagellates," *Journal of Cell Biology 24*: 387–400, 1965.
227. GRUNLOH, R. "Flying saucers," *Royal Anthropological Institute News 23*: 1–4, 1977.
228. GURNEY, E., et al. *Phantasms of the living.* London: Trubner, 1886.
229. GWYNNE, P. In *Newsweek*: New York, November 28, 1977.
230. HAILE, B. *Starlore among the Navaho.* Sante Fe: Gannon, 1977.
231. HALDANE, J. B. S. "The origins of life," *New Biology 16*: 12–27, 1954.
232. HALL, C. S., & NORDBY, V. J. *A primer of Jungian psychology.* New American Library: New York, 1973.
233. HALL-CRAGGS, J. "The development of song in the blackbird," *Ibis 104*: 277–300, 1962.
234. HALL-CRAGGS, J. "The aesthetic content of bird song," in HINDE (253).
235. HALLOWELL, A. I. "Cultural factors in the structuralization of perception," in ROHRER & SHERIF (460).
236. HALLOWELL, A. I. "Self, society, and culture in phylogenetic perspective," in TAX (531).
237. HAMBURG, D. A., et al. (eds.). *Perception and its disorders.* Baltimore: Williams & Wilkins, 1970.
238. HARADA, K., & FOX, S. W. "Thermal synthesis of natural amino acids . . ." *Nature 201*: 335–336, 1965.
239. HARRISON, M. *Fire from heaven.* London: Sidgwick & Jackson, 1976.
240. HARTMAN, W. D., & REISWEG, H. M. "The individuality of sponges," in BOARDMAN et al. (62).
241. HARTMANN, E. L. *The functions of sleep.* New Haven: Yale University Press, 1973.
242. HARTMANN, E. L. (ed.). *Sleep and dreaming.* Boston: Little, Brown, 1970.
243. HARWOOD, A. *Witchcraft, sorcery and social categories among the Safwa.* Oxford: Oxford University Press, 1970.

244. HAWKINS, D. R. "Psychoanalytic dream theory reexamined," in HARTMANN (242).
245. HAYNES, R. *The seeing eye, the seeing I.* London: Hutchinson, 1976.
246. HEISENBERG, W. *Physics and philosophy.* London: Allen & Unwin, 1963.
247. HELD, R., & BOSSOM, J. "Neonatal deprivation and adult rearrangement," *Journal of Comparative and Physiological Psychology 54*: 33–37, 1961.
248. HELD, R., & HEIN, A. "Movement produced stimulation in the development of visually guided behavior," *Journal of Comparative and Physiological Psychology 56*: 872–876, 1963.
249. HENRY, S. M. (ed.). *Symbiosis.* New York: Academic Press, 1966.
250. HERRNSTEIN, R. J., & LOVELAND, D. H. "Complex visual concept in the pigeon," *Science 146*: 549–551, 1964.
251. HILGARD, E. R. "A neo-dissociation theory of pain reduction in hypnosis," *Psychological Review 80*: 396–411, 1973.
252. HILLMAN, H. & SARTORY, P. "The unit membrane, the endoplasmic reticulum, and the nuclear pores are artefacts," *Perception 6*: 667–673, 1978.
253. HINDE, R. A. (ed.). *Bird vocalizations.* Cambridge: Cambridge University Press, 1969.
254. HINDLEY, K. "Tar among the stars," *New Scientist 76*: 23, 1977.
255. HINSHELWOOD, C. N. Presidential Address to British Association for the Advancement of Science, 1965.
256. HINTZE, N. A., & PRATT, J. G. *The psychic realm.* New York: Random House, 1975.
257. HIRSCH, A. *Handbook of geographical and historical pathology.* New Sydenham Society: London, 1886.
258. HITCHING, F. *Pendulum.* London: Fontana, 1977.
259. HOFFMAN, J. G. *The life and death of cells.* London: Hutchinson, 1958.
260. HOLLDOBLER, B. "Communication between ants and their guests," *Scientific American 224*: 86–93, 1971.
261. HOLLOWAY, R. L. "The casts of fossil hominid brains," *Scientific American 231*: 106–114, 1974.
262. HOLROYD, S. *Prelude to a landing on Earth.* London: W. H. Allen, 1977.
263. HOWELL, J. H. "The life cycle of the sea lamprey . . ." in SMITH (494).
264. HOYLE, F. *The black cloud.* London: Heinemann, 1960.
265. HOYLE, F., & WICKRAMASINGHE, N. C. "Primitive grain clumps and organic compounds in carbonaceous chondrites," *Nature 264*: 45–46, 1976.
266. HOYLE, F. & WICKRAMASINGHE, N. C. "Prebiotic molecules and intersellar grain clumps," *Nature 266*: 241–243, 1977.

267. HOLYE, F., & WICKRAMASINGHE, N. C. "Does epidemic disease come from space?" *New Scientist 76*: 402–404, 1977.

268. HOYLE, F., & WICKRAMASINGHE, N. C. "Origin and nature of carbonaceous material in the galaxy," *Nature 270*: 701–703, 1977.

269. HUBEL, D. H., & WIESEL, T. N. "Receptive fields, binocular interaction and functional architecture in the cat's visual cortex," *Journal of Physiology 160*: 106, 1962.

270. HUTCHISON, R. E., et al. "The basis for individual recognition by voice in the Sandwich tern," *Behaviour 32*: 150–157, 1968.

271. HUXLEY, A. *Plant and planet*. London: Allen Lane, 1974.

272. HUXLEY, J. S. *The individual in the animal kingdom*. Cambridge University Press: Cambridge, 1912.

273. HUXLEY, J. S. "The evolution of life," in TAX (531).

274. HUXLEY, J. S. "A discussion of ritualization of behaviour in animals and man," *Philosophic Proceedings of the Royal Society B271*, 1966.

275. HYNEK, J. A. *The UFO experience*. London: Abelard Schuman, 1972.

276. IKEMA, Y., & NAKAGAWA, S. "A psychosomatic study of contagious dermatitis," *Kyushu Journal of Medical Science 13*: 335–350, 1962.

277. IMANISHI, K. "Social behaviour in Japanese monkeys," *Psychologia 1*: 47–54, 1957.

278. IMMS, A. D. *Insect natural history*. London: New Naturalist, 1947.

279. INGLIS, B. *Natural and supernatural*. London: Hodder & Stoughton, 1978.

280. ITARD, J.-M. G. *De l'education du jeune sauvage de l'Aveyron*. Paris: Imprimerie Impériale, 1801.

281. IVERSON, J. *More lives than one*. London: Souvenir, 1976.

282. JACOBI, J. *The psychology of C. G. Jung*. London: Routledge & Kegan Paul, 1951.

283. JASTROW, J. *Error and eccentricity*. New York: Dover, 1962.

284. JAYNES, J. *The origin of consciousness in the breakdown of the bicameral mind*. Boston: Houghton Mifflin, 1976.

285. JENNINGS, H. S. *Behaviour of lower organisms*. New York: Columbia University Press, 1906.

286. JENNINGS, D. H., & LEE, D. L. (eds.). "Symbiosis," *Symposia of the Society for Experimental Biology 29*, 1975.

287. JEON, K. W., & DANIELLI, J. F. "Microsurgical studies with large free living amoebas," *International Reviews of Cytology 30*: 49–89, 1971.

288. JOHN, P., & WHATLEY, F. R. "*Paracoccus dentrificans . . .*" in JENNINGS & LEE (286).

289. JONAS, D., & KLEIN, D. *Man-child*. London: Jonathan Cape, 1971.

290. JONES, H. S., & OSWALD, I. "Two cases of healthy insomnia,"

*Journal of Electroencephalography and Clinical Neurophysiology 24;* 378–380, 1968.

291. JOSEPHSON, B. "Possible connections betwen psychic phenomena and quantum mechanics," *New Horizons 2:* 224–226, 1975.

292. JOSEPHSON, R. K., & MACKIE, G. O. "Multiple pacemakers and the behavior of the hydroid *Tubularia*," *Journal of Experimental Biology 43:* 293–332, 1965.

293. JUNG, C. G. *The role of the unconscious.* (1918) Collected Works *10.* London: Routledge & Kegan Paul.

294. JUNG, C. G. *The relations between the ego and the unconscious.* (1928) Collected Works 7.

295. JUNG, C. G. *The concept of the collective unconscious.* (1936) Collected Works *9.*

296. JUNG, C. G. *Individual dream symbolism in relation to Alchemy.* (1936) Collected Works *12.*

297. JUNG, C. G. *The fight with the Shadow.* (1946) Collected Works *10.*

298. JUNG, C. G. *Epilogue to essays on contemporary events.* (1947) Collected Works *10.*

299. JUNG, C. G. *Synchronicity: An acausal connecting principle.* (1952) Collected Works *8.*

300. JUNG, C. G. *Flying saucers: A modern myth of things seen in the skies.* (1959) Collected Works *10.*

301. JUNG, C. G. *Memories, dreams, reflections.* New York: Random House, 1961.

302. JUNG, C. G. *Analytical psychology.* London: Routledge & Kegan Paul, 1968.

303. KALES, A., et al. "Somnambulism," *Archives of General Psychiatry 14:* 595–604, 1966.

304. KANT, I. *Critique of practical reason.* Chicago: University of Chicago Press, 1949.

305. KAPLAN, E. A. "Hypnosis and pain," *Archives of General Psychiatry 2:* 567–568, 1960.

306. KARAKASHIAN, M. W. "Symbiosis in *Paramecium bursaria,*" in JENNINGS & LEE (286).

307. KARAKASHIAN, S. J., & SIEGEL, R. W. "A genetic approach to endocellular symbioses," *Experimental Parasitology 17:* 103–122, 1965.

308. KARLSON, P., & BUTENANDT, A. "Pheromones in insects," *Annual Review of Entomology 4:* 39–58. 1959.

309. KASTLE, W. "Soziale Verhaltensweisen von Chamäleonen," *Zietschrift für Tierpsychologie 24:* 313–341, 1967.

310. KAWAI, M. "On the newly acquired behaviours of the natural troop of Japanese monkeys on Koshima Island," *Primates 4:* 113–115, 1963.

311. KAWAI, M. "Newly acquired precultural behaviour of the natural troop of Japanese monkeys on Koshima Island," *Primates 6:* 1 to 30, 1965.

312. KAWAMURA, S. "The process of sub-cultural propagation among Japanese monkeys," in SOUTHWICK (497).

313. KEEL, J. A. *Strange creatures from time and space*. London: Neville Spearman, 1975.

314. KEITH, A. *A new theory of human evolution*. London: Watts & Co., 1948.

315. KENNEDY, J. E., & TADDONIO, J. L. "Experimenter effects in parapsychological research," *Journal of Parapsychology 40*: 1–33, 1976.

316. KLINE, M. V. "Hypnosis and age progression." *Journal of Genetical Psychology 78*: 195–206, 1951.

317. KNACKE, R. F. "Carbonaceous compounds in interstellar dust," *Nature 269*: 132–133, 1977.

318. KNOX, V. J., et al. "Pain and suffering in ischemia," *Archives of General Psychiatry 30*: 840–847, 1974.

319. KOCH, A. "Intracellular symbiosis in insects," *Annual Review of Microbiology 14*: 121–140, 1960.

320. KOESTLER, A. *The ghost in the machine*. London: Hutchinson, 1967.

321. KOESTLER, A. *The roots of coincidence*. London: Hutchinson, 1972.

322. KOESTLER, A. *Janus*. London: Hutchinson, 1978.

323. KOESTLER, A., & SMYTHIES, J. R. (eds.). *Beyond reductionism*. London: Hutchinson, 1969.

324. KONIJN, T. M., et al. "The acrasin activity of adenosine—3', 5'—cyclic phosphate," *Proceedings of the National Academy of Sciences 58*: 1152–1154, 1967.

325. KRANTZ, G. S. "Pithecanthropine brain size and its culutral consequences," *Man 11*: 85–87, 1961.

326. KRANTZ, G. S. "Brain size and hunting ability in earliest man," *Current Anthropology 9*: 450–451, 1968.

327. KRIPPNERS, S. (ed.). *Advances in parapsychological research psychokinesis*. New York: Plenum Press, 1977.

328. KUBIE, L. S. "Illusion nad reality in the study of sleep . . ." *International Journal of Clinical and Experimental Hypnosis 20*: 205–223, 1972.

329. KUBIE, L. S., & MARGOLIN, S. "An apparatus for the use of breath sounds as a hypnagogic stimulus," *American Journal of Psychiatry 100*: 610, 1944.

330. LACK, D. "The behaviour of the robin," *Proceedings of the Zoological Society of London 109*: 169–178, 1939.

331. LAMB, I. M. "Lichens," *Scientific American 201*: 144–156, 1959.

332. LANGEN, D. "Peripheral changes in blood circulation during autogenic training and hypnosis," in CHERTOK (106).

333. LANGER, S. K. *Feeling and form*. New York: Scribner's, 1953.

334. LANGER, S. K. *Mind: An essay on human feeling* (2 vols.). Baltimore: Johns Hopkins University Press, 1972.

335. LANNERS, E. *Illusions*. London: Thames & Hudson, 1977.
336. LASSNER, J. (ed.). *Hypnosis and psychosomatic medicine*. Berlin: Springer, 1967.
337. LAURIE, P. "More about knowledge," *New Scientist 62*: 248–249, 1974.
338. LEAKEY, R. E., & LEWIN, R. *Origins*. London: Macdonald & Jane's, 1977.
339. LEDOUX, L., et al. "DNA mediated genetic correction of thiamineless *Arabidopsis thaliana*," *Nature 249*: 17–21, 1974.
340. LEE, D. *Freedom and culture*. Englewood Cliffs, N.J.: Prentice-Hall, 1959.
341. LEE, S. G. M., & MAYES, A. R. (eds.). *Dreams and dreaming*. Harmondsworth, England: Penguin, 1973.
342. LEIDENFROST, J. G. "On the fixation of water in diverse fire" (1756), *International Journal of Heat and Mass Transfer 9*: 1153–1166, 1966.
343. LERNER, B. "Dream function reconsidered," *Journal of Abnormal Psychology 72*: 85–100, 1966.
344. LEVINSON, B. W. "States of awareness during general anesthesia," in LASSNER (336).
345. LEWIS, H. B. "The royal road to the unconscious," in HARTMANN (242).
346. LIBERMAN, A. M. "Some characteristics of perception in the speech mode," in HAMBURG et al. (237).
347. LILLY, J. *Centre of the cyclone*. St. Albans: Paladin, 1973.
348. LINDEN, E. *Apes, men and language*. New York: E. P. Dutton, 1975.
349. LINDSEY, H., & CARLSON, C. C. *The late, great planet Earth*. Grand Rapids: Zondervan, 1970.
350. LINN, E. L. "Verbal auditory hallucinations," *Journal of Nervous and Mental Disease 164*: 8–17, 1977.
351. LLAURADO, J. G., et al. (eds.). *Biological and clinical effects of low frequency magnetic and electrical fields*. Springfield, Ill.: Charles Thomas, 1974.
352. LORENZ, K. "Der Kumpan in der Umwelt des Vogels," *Journal of Ornithology 83*: 137–413, 1935.
353. LORENZ, K. "The mind of man," in LANNERS (335).
354. LOVELL, A. C. B. "Meteors," in BATES (37).
355. LUDWIG, A. M., et al. "The objective study of a multiple personality," *Archives of General Psychiatry 26*: 298–310, 1972.
356. LURIA, A. R., & YUDOVICH, F. *Speech and development of higher psychological functions in the child*. New York: Stagler, 1959.
357. MAETERLINCK, M. *The unknown guest*. London: Methuen, 1914.
358. MAGOUN, H. W. *The waking brain*. Springfield, Ill.: Charles Thomas, 1963.

359. MANNING, M. *The link*. Gerrards Cross: Colin Smythe, 1974.
360. MARAIS, E. *The soul of the white ant*. London: Jonathan Cape, 1971.
361. MARCUSE, F. L. *Hypnosis*. Harmondsworth, England: Penguin, 1959.
362. MARGULIS, L. *Origin of eukaryotic cells*. New Haven: Yale University Press, 1970.
363. MARGULIS, L. "Symbiosis and evolution," *Scientific American* 225: 48–57, 1971.
364. MARGULIS, L. "Five kingdom classification and the origin and evolution of cells," *Evolutionary Biology* 7: 45–78, 1974.
365. MARGULIS, L. "Genetic and evolutionary consequences of symbiosis," *Experimental Parasitology 39*: 277–349, 1976.
366. MARLER, P. "Comparative study of song development in Emberizine finches," *Proceedings of the Fourteenth International Ornithological Congress*: 231–244, 1967.
367. MARSHALL, A. J. *Bowerbirds, their displays and breeding cycles*. Oxford: Oxford University Press, 1954.
368. MASLOW, A. *Toward a psychology of being*. Princeton, N.J.: Van Nostrand, 1962.
369. MASON, A. A. "A case of congenital ichthyosiform . . ." *British Medical Journal* (2): 422–423, 1952.
370. MATTHEWS, H. "Hibernation and sleep," in WOLSTENHOLME (598).
371. MEDDIS, R. "On the function of sleep," *Animal Behaviour 23*: 676–691, 1975.
372. MEDDIS, R., et al. "An extreme case of healthy insomnia," *Electroencephalography and Clinical Neurophysiology 35*: 213–214, 1973.
373. MEHRA, J. (ed.). *The physicist's conception of nature*. Dordrecht: Reidel, 1973.
374. MICHEL, A. *Flying saucers and the straight line mystery*. New York: Criterion, 1958.
375. MICHELL, J. *The Earth spirit*. London: Thames & Hudson, 1975.
376. MICHELL, J., & RICKARD, R. J. M. *Phenomena*. London: Thames & Hudson, 1977.
377. MILLER, S. L. "A production of amino acids under possible primitive earth conditions," *Science 117*: 528–529, 1953.
378. MILLER, S. L., & ORGEL, L. E. *The origins of life on the earth*. Englewood Cliffs, N.J., Prentice-Hall, 1974.
379. MISHLOVE, J. *The roots of consciousness*. New York: Random House, 1975.
380. MONOD, J. L. *Chance and necessity*. London: Collins, 1972.
381. MONOD, J. L. "On the molecular theory of evolution," in AYALA (24).
382. MONTAGUE, A. "Neoteny and the evolution of the human mind," *Explorations 6*: 85–90, 1956.
383. MOONEY, J. "The Ghost-Dance religion and the Sioux out-

break of 1890," *Annual Report of the Bureau of American Ethnology 14*: 641–1136, 1896.

384. MORGAN, B. *Men and discoveries in mathematics.* London; John Murray, 1972.

385. MORRIS, D. (ed.). *Primate ethology.* New York: Doubleday, 1969.

386. MORRIS, J. D., et al. *Research in parapsychology.* Metuchen, N.J.: Scarecrow, 1975.

387. MORUZZI, G., & MAGOUN, H. W. "Brain stem reticular formation and activation of the EEG," *Electroencephalography and Clinical Neurophysiology 1*: 455–473, 1949.

388. MULLINS, J. F., et al. "Pachyonychia congenita treated by hypnosis," *Archives of Dermatology 71*: 264, 1955.

389. MYERS, F. W. H. *Human personality and its survival of bodily death.* London: Longmans, 1903.

390. McCALL, T. S., & LEVITT, Z. *Satan in the sanctuary.* Chicago: Moody, 1973.

391. McCAMPBELL, J. M. *Ufology.* Millbrae, Calif.: Celestial Arts, 1976.

392. McINTYRE, J. *Mind in the waters.* New York: Scribner's, 1974.

393. McKELLAR, P. *Imagination and thinking.* London: Cohen & West, 1957

394. MACKIE, G. O. "Analysis of locomotion in a siphonophore colony," *Proceedings of the Royal Society B 159*: 366–391, 1964.

395. MACKINTOSH, J. H., & GRANT, E. C. "The effect of olfactory stimuli on the agonistic behaviour of laboratory mice," *Zeitschrift für Tierpsychologie 23*: 584–587, 1966.

396. McKINNEY, F. "The comfort movements of Anatidae," *Behaviour 25*: 120–220, 1965.

397. MACLEAN, N. *The differentiation of cells.* London: Edward Arnold, 1977.

398. MACLEAN, P. D. "A triune concept of the brain and behaviour," in BOAG & CAMPBELL (61).

399. McNEILL, W. H. *Plagues and peoples.* Oxford: Blackwell, 1976.

400. NAPIER, J. *Bigfoot.* London: Jonathan Cape, 1972.

401. NATSOULAS, T. "Consciousness . . ." *Journal for the Theory of Social Behaviour 7*: 29–40, 1977.

402. NELSON-REES, W. A. & FLANDERMEYER, R. R. "Helen Lane cultures defined," *Science 191*: 96–98, 1975.

403. NELSON-REES, W. A., et al. "Banded marker chromosomes as indicators of intraspecies cellular contamination," *Science 184*: 1093–1096, 1974.

404. NEUMANN, E. *Art and the creative unconscious.* Princeton, N.J.: Princeton University Press, 1959.

405. NEWELL, P. C. "How cells communicate," *Endeavour 1*: 63–68, 1977.

406. NEWSON, E. Introduction to "Nadia," in SELFE (483).

407. NEWTON, R. *The Crime of Claudius Ptolemy.* Baltimore: Johns Hopkins, 1977.
408. NOVIKOFF, A. B., & HOLTZMAN, E. *Cells and organelles.* New York: Holt, Rinehart & Winston, 1976.
409. O'NEILL, J. *Prodigal genius.* New York: Washburn, 1944.
410. OPARIN, A. I. *The origin of life on the earth.* Edinburgh; Oliver and Boyd, 1957.
411. OPARIN, A. I. *Genesis and evolutionary development of life.* London: Academic Press, 1968.
412. ORNE, M. T. "Hypnotically induced hallucinations," in WEST (582).
413. ORNSTEIN, R. *The psychology of consciousness.* London: Jonathan Cape, 1975.
414. ORNSTEIN, R. *The mind field.* New York; Viking Press, 1976.
415. OSTY, E. *Supernormal faculties in man.* New York; E.P. Dutton, 1922.
416. OSWALD, I. *Sleep.* Harmondsworth, England, 1966.
417. OTERI, I. (ed.). *Quantum Physics and Parapsychology.*
418. OWEN, I. M., & SPARROW, M. *Conjuring up Philip.* Toronto: Fitzhenry & Whiteside, 1976.
419. PACKARD, V. *The people shapers.* London: Macdonald & Jane's, 1978.
420. PAGE. E. W., et al. *Human reproduction.* Philadelphia: Saunders, 1976.
421. PANTIN, C. F. A., & THORPE, W. H. (eds.). *Relations between the sciences.* Cambridge: Cambridge University Press, 1968.
422. PARKER, G. A., et al. "The origin and evolution of gamete dimorphism . . ." *Journal of Theoretical Biology 36:* 529–553, 1972.
423. PASTAN, I. "Cyclic AMP," *Scientific American 227:* 97–105, 1972.
424. PATTERSON, T. *Spirit photography.* London: Regency, 1965.
425. PATTIE, F. A. "The production of blisters by hypnotic suggestion," *Journal of Abnormal and Social Psychology 36:* 62–72, 1941.
426. PAUWELS, L., & BERGIER, J. *The dawn of magic.* London: Anthony Gibbs, 1963.
427. PEARCE, J. C. *The crack in the cosmic egg.* New York: Julian, 1971.
428. PEARCE, J. C. *Exploring the crack in the cosmic egg.* New York: Julian, 1974.
429. PENFIELD, W. "Speech, perception and the uncommitted cortex," in ECCLES (144).
430. PERUTZ, M. "Bizarre behaviour among the messengers," *New Scientist 77:* 8–9, 1978.
431. PETRUNKEVITCH, A. "Tarantula versus tarantula-hawk . . ." *Journal of Experimental Zoology 45:* 367–397, 1926.
432. PIRIE, N. W. "Chemical diversity and the origin of life," in CLARK (110).

433. PLAYFAIR, G. *The flying cow*. London: Souvenir, 1975.
434. POLANYI, M. *The tacit dimension*. New York: Doubleday, 1966.
435. PONNAMPERUMA, C. "Abiological synthesis of some nucleic acid constituents," in Fox (183).
436. POPPER, K. *Conjectures and refutations*. London: Routledge & Kegan Paul, 1963.
437. POPPER, K. *Objective knowledge*. Oxford: Oxford University Press, 1972.
438. POTTER, K., & ECCLES, J. *The self and its brain*. Berlin: Springer, 1977.
439. POTTER, S. *The theory and practice of gamesmanship*. London: Rupert Hart-Davis, 1947.
440. POULTON, E. B. "The terrifying appearance of *Laternaria . . .*" *Proceedings of the Royal Entomological Society of London 43*: 43, 1924.
441. PRATT, J. G. "Testing for an ESP factor in pigeon homing," in WOLSTENHOLME & MILLAR (579).
442. PRINCE, M. *The dissociation of a personality*. London: Longmans Green, 1905.
443. PRINCE, W. F. *The case of Patience Worth*. (1924) New York: University Books, 1964.
444. PUGH, G. E. *The biological origin of human values*. New York: Basic Books, 1977.
445. RANDALL, J. L. *Parapsychology and the nature of life*. London: Souvenir, 1975.
446. RAPPAPORT, R. A. *Pigs for the ancestors*. New Haven: Yale University Press, 1968.
447. RAPPAPORT, R. A. "The sacred in human evolution," *Annual Review of Ecology and Systematics 2*: 23–44, 1971.
448. REED, G. *The psychology of anomalous experience*. London: Hutchinson, 1972.
449. REICHEL-DOLMATOFF, G. *Amazonian cosmos*. Chicago: University of Chicago Press, 1971.
430. REID, C., & ORGEL, L. E. "Synthesis of sugars in potentially prebiotic conditions," *Nature 216*: 455, 1967.
451. RENARD, J. *The journal of Jules Renard*. (1906) New York: Braziller, 1964.
452. RHINE, J. B., & FEATHER, S. R. "The study of cases of psi-trailing in animals," *Journal of Parapsychology 26*: 1–22, 1962.
453. ROBERTS, J. *The Seth material*. Englewood Cliffs, N.J.: Prentice-Hall, 1970.
454. ROBERTS, J. *Seth speaks*. Englewood Cliffs, N.J.: Prentice-Hall, 1971.
455. ROBERTS, J. *The nature of personal reality*. Englewood Cliffs, N.J.: Prentice-Hall, 1972.
456. ROFFWARD, P., et al. "Ontogenetic development of the human sleep-dream cycle," *Science 152*: 604–619, 1966.
457. ROGERS, J. "Genes in pieces," *New Scientist 77*: 18–20, 1978.

458. ROGERS, M. "The double-edged helix," *Rolling Stone*, March 25, 1976.
459. ROHEIM, G. *Psychoanalysis and anthropology.* New York; International Universities Press, 1950.
460. ROHRER, J. H., & SHERIF, M. (eds.). *Social psychology at the crossroads.* New York: Harper, 1951.
461. ROLDAN, E., et al. "Excitability changes during the sleep cycle of the rat," *Electroncephalographic and Clinical Neurophysiology 15*: 775–785, 1963.
462. ROLL, W. G. Comments on STEVENSON (511).
463. ROODYN, D. B., & WILKIE, D. *The biogenesis of mitochondria.* London: Methuen, 1968.
464. ROSE, S. *The chemistry of life.* Harmondsworth, England: Penguin, 1970.
465. ROSEBURY, T. *Life on man.* London: Secker & Warburg, 1969.
466. ROSLANSKY, J. D. (ed.). *The uniqueness of man.* Amsterdam: North-Holland, 1969.
467. RUBIN, H. "A defective cancer virus," *Scientific American 210*: 46–52, 1964.
468. RUBINSTEIN, R., & NEWMAN, R. "The living out of future experiences under hypnosis," *Science 119*: 472–473, 1954.
469. RUSH, J. H. "Problems and methods in psychokinesis research," in KRIPPNER (327).
470. SAGAN, C. "Interstellar organic chemistry," *Nature 238*: 77–80, 1972
471. SAGAN, C. *The dragons of Eden.* New York: Random House, 1977.
472. SAKATA, A., et al. "Spectroscopic evidence for interstellar grain clumps in meteoritic inclusions," *Nature 266*: 241, 1977.
473. SAMUEL, Lord. *Belief and action.* London: Crescent, 1947.
474. SANDERSON, I. *Uninvited visitors.* London: Neville Spearman, 1969.
475. SCHENKEL, R. "Zur Deutung der Phasianidenbalz," *Ornithologica Beobachtungen 53*: 182, 1956.
476. SCHMIDT, H. "PK experiments with animals as subjects," *Journal of Parapsychology 34*: 255–261, 1970.
477. SCHOUTEN, S. A. "Psi in mice . . ." *Journal of Parapsychology 36*: 261–282, 1972.
478. SCHREIBER, F. R. *Sybil.* London: Allen Lane, 1974.
479. SCHRIER, A., & STOLLNIZ, F. (eds.). *Behavior of non-human primates.* Academic Press: New York, 1971.
480. SEASHORE, C. E. "Measurements of illusions and hallucinations in normal life," *Studies of Yale Psychology Laboratory 2*: 1–67, 1895.
481. SEBEOK, T. A. (ed.). *Animal communication.* Bloomington: Indiana University Press, 1968.
482. SEGALL, M. H., et al. *The influence of culture on visual perception.* Indianapolis: Bobbs-Merrill, 1966.

483. Selfe, L. *Nadia*. London: Academic Press, 1977.
484. Sheehan, P. W., & Perry, C. W. *Methodologies of hypnosis*. New Jersey: Lawrence Erlbaum, 1976.
485. Sheer, D. E. (ed.). *Electrical stimulations of the brain*. Austin: University of Texas Press, 1961.
486. Sherrington, C. *Man on his nature*. Cambridge: Cambridge University Press, 1940.
487. Shklovskii, I. S., & Sagan, C. *Intelligent life in the universe*. London: Picador, 1977.
488. Simpson, T. L. "Coloniality among the Porifera," in Boardman (62).
489. Singer, J. L. *Daydreaming and fantasy*. London: Allen & Unwin, 1976.
490. Singh, J. *Modern cosmology*. Harmondsworth, England: Penguin, 1970.
491. Smith, A. *The body*. London: Allen & Unwin, 1968.
492. Smith, J. M. "Evolution and the theory of games," *American Scientist 64*: 41–45, 1976.
493. Smith, J. M. "The limitations of evolution theory," in Duncan (141).
494. Smith, R. T., et al. *Phylogeny of immunity*. Gainesville: University of Florida Press, 1966.
495. Snyder, F. "Toward an evolutionary theory of dreaming," *American Journal of Psychiatry 123*: 121–142, 1966.
496. Sondheimer, E., & Simeone, J. B. (eds.). *Chemical ecology*. Academic Press: New York, 1970.
497. Southwick, C. H. *Primate social behavior*. Princeton, N.J.: Van Nostrand, 1963.
498. Spemann, H. *Embryonic development and induction*. New Haven: Yale University Press, 1938.
499. Spencer, W. B., & Gillen, F. J. *The Arunta*. London: Macmillan, 1927.
500. Sperry, R. W. "Problems outstanding in the evolution of brain function," James Arthur Lecture, American Museum of Natural History, 1964.
501. Sperry, R. W. "The great cerebral commissure," *Scientific American 210*: 42–52, 1964.
502. Sperry, R. W. In Duncan & Weston-Smith (141).
503. Stanford, R. *Socorro saucer in a Pentagon pantry*. Austin: Blue Apple. 1976.
504. Stanford, R. G. "An experimentally testable model for spontaneous psi events," *Journal of the American Society for Psychical Research 68*: 34–57 and 321–356, 1974.
505. Stanford, R. G., et al. "Psychokinesis as a psi-mediated instrumental response," *Journal of the American Society for Psychical Research 69*: 127–134, 1975.
506. Stayton, D. J., et al. "Infant obedience and maternal behaviour," *Child Development 42*: 1057–1069, 1971.

507. STEVENSON, I. "The evidence for survival from claimed memories of former incarnations," *Journal of the American Society for Psychical Research 54*: 51–71, 1960.

508. STEVENSON, I. *Twenty cases suggestive of reincarnation*. Charlottesville: University Press of Virginia, 1974.

509. STEVENSON, I. *Xenoglossy*. Charlottesville: University Press of Virginia, 1974.

510. STEVENSON, I. "The explanatory value of the idea of reincarnation," *Journal of Nervous and Mental Disease 164*: 305–326, 1977.

511. STEVENSON, I. "Research into the evidence of man's survival after death," *Journal of Nervous and Mental Disease 165*: 152–170, 1977.

512. STEVENSON, J., et al. "Individual recognition by auditory cues in the common term," *Nature 226*: 562–563, 1970.

513. STEWARD, F. C. *Growth and organization in plants*. Reading: Addinson-Wesley, 1968.

514. STOCKS, P. "A study of cancer mortality . . ." *British Journal of Cancer 15*: 701–711, 1961.

515. STOCKS, P., & DAVIES, R. J. "Epidemiological evidence from chemical and spectrograpic analyses that soil is concerned in the causation of cancer," *British Journal of Cancer 14*: 8–22, 1960.

516. STORY, R. *The space gods revealed*. London: New English Library, 1976.

517. STRINGER, E. T. *The secret of the gods*. London: Neville Spearman; 1974.

518. STROMEYER, C. F., & PSOTKA, J. "The detailed texture of eidetic images," *Nature 225*: 346–349, 1970.

519. STROUN, M., et al. "Natural release of nucleic acids from bacteria into plant cells," *Nature 227*: 607–608, 1970.

520. SUMMERS, R. *Ancient mining in Rhodesia*. Memoir No. 3, National Museum, Salisbury, 1969.

521. SWANN, I. *To kiss Earth goodbye*. New York: Hawthorn, 1975.

522. TANSLEY, D. *Omens of awareness*. Suffolk: Neville Spearman, 1977.

523. TART, C. T. "Physiological correlates of psi cognition," *International Journal of Parapsychology 5*: 375–386, 1963.

524. TART, C. T. "Psychedelic experience associated with a novel hypnotic procedure, mutual hypnosis," in TART (528).

525. TART, C. T. *States of consciousness*. New York: E. P. Dutton, 1975.

526. TART, C. T. *Learning to use extrasensory perception*. Chicago: University of Chicago Press, 1976.

527. TART, C. T. *Psi*. New York: E. P. Dutton, 1977.

528. TART, C. T. (ed.). *Altered states of consciousness*. New York: Doubleday, 1972.

529. TAUBER, E. S., et al. "Eye movements and EEG activity during sleep in diurnal lizards." *Nature 212*: 1612–1613, 1966.
530. TAUBER, E. S., & WEITZMAN, E. D. "Eye movements during behavioural inactivity in certain Bermuda reef fish," *Psychophysiology 6*: 230, 1969.
531. TAX, S. (ed.). *Evolution after Darwin*. Chicago; University of Chicago Press, 1960.
532. TAYLOR, F. J. R. "Autogenous theories for the origin of eukaryotes," *Taxton 25*: 377–390, 1976.
533. TAYLOR, G. R. *The science of life*. London: Thames and Hudson, 1963.
534. THÉODOR, J. L. "Distinction between self and not-self in lower invertebrates," *Nature 227*: 690–692, 1970.
535. THIELCKE, G. "Geographic variation in bird vocalization," in HINDE (253).
536. THOMAS, L. *The lives of a cell*. New York: Viking Press, 1974.
537. THORPE, W. H. "Vitalism and organicism," in ROSLANSKY (466).
538. THORPE, W. H. *Animal nature and human nature*. London: Methuen, 1974.
539. THORPE, W. H. *Biology and the nature of man*. Oxford: Oxford University Press, 1962.
540. THORPE, W. H. & NORTH, M. E. W. "Vocal imitation in the tropical boubou shrike," *Ibis 108*: 432–435, 1966.
541. TIMOURIAN, H. "Symbiotic emergence of metazoans," *Nature 226*: 283–284, 1970.
542. TINBERGEN, N. "Social releasers and the experimental method required for their study," *Wilson Bulletin 60*: 6–52, 1948.
543. TINBERGEN, N. *The study of instinct*. Oxford; Oxford University Press, 1951.
544. TOMPKINS, P., & BIRD, C. *The secret life of plants*. New York: Harper & Row, 1973.
545. TRENCH, R. "Of leaves that crawl . . ." in JENNINGS & LEE (286).
546. TRIVERS, R. L. "Parental investment and sexual selection," in CAMPBELL (97).
547. TRUE, R. "Experimental control in hypnotic age regression states," *Science 110*: 583–584, 1949.
548. TSUMORI, A. "Newly acquired behaviour and social interaction of Japanese monkeys," in ALTMANN (5).
549. TYLER, H. A. *Pueblo gods and myths*. Norman: University of Oklahoma Press, 1964.
550. TYRRELL, G. N. M. *Apparitions*. London: Duckworth, 1943.
551. ULLMAN, M. "Parapsychology and psychiatry," in FREEDMAN (186).
552. ULLMAN, M., et al. *Dream telepathy*. London: Turnstone, 1973.
553. UNGAR, G., et al. "Isolation, identification and synthesis of a

specific-behaviour-inducing brain peptide," *Nature 238*: 198–202, 1972.

554. Vallée, J. *Anatomy of a phenomenon.* London: Neville Spearman, 1966.
555. Vallée, J. *Passport to Magonia.* London: Neville Spearman, 1970.
556. Vallée, J. *UFOs: The psychic solution.* St. Albans: Panther, 1977.
557. Vallée, J., & Vallée, J. *Challenge to science.* London: Neville Spearman, 1967.
558. Van Olphen, H. *An introduction to clay colloid chemistry.* New York: Interscience Publishers, 1963.
559. Van Twyver, H., & Allison, T. "EEG study of the shrew," *Psychophysiology 6*: 231, 1969.
560. Van Valen, L. "A new evolutionary law," *Evolution Theory 1*: 1–30, 1973.
561. Vernon, J. *Inside the black room.* London: Souvenir, 1965.
562. Verwey, J. "Die Paarungsbiologie des Fischreihers," *Zoologische Jahrbucher 48*: 1–120, 1930.
563. Von Bertalanffy, L. *Problems of life.* New York: Harper, 1960.
564. Von Däniken, E. *Chariots of the gods?* London: Souvenir, 1969.
565. Waddington, G. H. (ed.). *Towards a theoretical biology.* Edinburgh University Press: Edinburgh, 1969.
566. Wagner, J. C. "Experimental production of mesothelial tumours . . ." *Nature 196*: 180–181, 1962.
567. Walker, J. Jr. *Scientific American 237*: 126–131, 1977.
568. Wall, P. D. "Why do we not understand pain," in Duncan (141).
569. Washburn, S. L. "The evolution of human behaviour," in Roslansky (466).
570. Watson, L. *Omnivore.* London: Souvenir, 1971.
571. Watson, L. *Supernature.* London: Hodder & Stoughton, 1973.
572. Watson, L. *The Romeo error.* London: Hodder & Stoughton, 1974.
573. Watson, L. *Gifts of unknown things.* London: Hodder & Stoughton, 1976.
574. Webb, W. B. *Sleep: The gentle tyrant.* Englewood Cliffs, N.J.: Prentice-Hall, 1975.
575. Weinberg, S. *The first three minutes.* London: André Deutsch, 1977.
576. Weldon, J., & Levitt, Z. *UFOs: What on Earth is happening?* Irvine, Calif. Harvest House, 1975.
577. Weiss, A. Contribution to the 1961 Tenth Conference on Clays and Clay Mineralogy, in Cairns-Smith (91).
578. Weiss, P. "The living system," in Koestler & Smythies (323.).
579. Werner, M. W. "The relationship of interstellar molecules to the origin of life," *Icarus 15*: 325–355, 1971.

580. WEST, L. J. *Hallucinations.* New York: Grune & Stratton, 1962.
581. WHEELER, J. A. In MEHRA, J. (373).
582. WHEELER, W. M. "The ant colony as an organism," *Journal of Morphology 22:* 307–325, 1911.
583. WHITE, L. "Four stages in the evolution of minding," in TAX (531).
584. WHITE, R. A. "The limits of experimenter influence in psi test results," *Journal of the American Society for Psychical Research 70:* 333–370, 1976.
585. WHITEHEAD, A. N. *Adventures of ideas.* San Francisco: Free Press, 1967.
586. WHITTON, J. L. "Qualitative time domain analysis of acoustic envelopes of psychokinetic table rappings," *New Horizons 2:* 21–24, 1975.
587. WICKLER, W. "Socio-sexual signals . . ." in MORRIS (385).
588. WICKRAMASINGHE, N. C. "Where life begins?" *New Scientist 74:* 119–121, 1977.
589. WILLIAMS, G. C. *Sex and evolution.* Princeton, N.J.: Princeton University Press, 1975.
590. WILLIAMS, H. L., et al. "Impaired performance with acute sleep loss," *Psychological Monographs 73:* 14–78, 1959.
591. WILSON, E. B. *The cell in development and heredity.* Macmillan: New York, 1975.
592. WILSON, E. O. *The insect societies.* Cambridge, Mass.: Harvard University Press, 1971.
593. WILSON, E. O. *Sociobiology: The new synthesis.* Cambridge, Mass.: Harvard University Press, 1975.
594. WOLMAN, Y., et al. "Non protein amino acids from spark discharges . . ." *Proceedings of the National Academy of Sciences 69:* 809–811, 1972.
595. WOLSTENHOLME, G. E. W., & MILLAR, E. C. P. (eds.). *Extrasensory perception.* New York: Citadel, 1956.
596. WOLSTENHOLME, G. E. W., & CONNOR, M. (eds.). *The nature of sleep.* Boston: Little, Brown, 1960.
597. WORTIS, J. (ed.). *Recent advances in biological psychiatry.* New York: Plenum, 1965.
598. YEATS, W. B. *A vision.* London: Macmillan, 1938.
599. ZIMMER, H. *Myths and symbols in Indian art and civilization.* Princeton, N.J.: Princeton University Press, 1972.
600. ZINSSER, H. *Rats, lice and history.* London: Jonathan Cape, 1935.

With this premise, Dr. Lyall Watson, the gifted and perceptive author of <u>Supernature</u>, enters into a breathtaking examination of the origins and evolution of the universe and our role within it that spans the traditional boundaries of the sciences.

Taking a brilliantly eclectic approach to his subject, Dr. Watson suggests that we are at the verge of a new wave of thought and interpretation, a critical stage in our understanding of reality and of ourselves, similar to the revolutionary waves brought about by Darwin and Freud—but vaster and more significant, because its source is in a general rise in our collective consciousness. In <u>Lifetide</u>, he looks at this development in our awareness in evolutionary terms, and finds that our whole conception of ourselves and our origins may soon come into question and be replaced by an entirely different view, an understanding that will take into account what Freud called "the black tide" of occultism, the mysterious and hidden forces that shape life in all its miraculous forms.

With the exception of a small minority that includes Carl Sagan and Fred Hoyle, modern scientists are a practical and industrious group, going on with their research in a careful, methodical manner but rarely pausing to consider the wonder of their discoveries. Lyall Watson sets out to bridge this gap in our understanding; to draw the linkages between the recent discoveries in astronomy, biology, and psychology into a unified picture of the cosmos. He tells of his personal experiences with "magic," recounting how a five-year-old girl defied the laws of physics by turning a tennis ball inside out before his eyes, and describes how the seeds of life itself are being carried through organic molecules in interstellar space. He finds new meanings in recent scientific developments, seeing in the discovery that viruses are able to transfer genetic information between species a vehicle for a whole new theory of evolution, and investigates the mysteries of cancer, relating the extraordinary story of Helen Lane, whose tissue culture (now called HeLa) continues to